2010-2011 Inventor's Market

Where to Sell or License Your Ideas, Products & Inventions

Julie Momyer

Cover photography courtesy of Renjith Krishnan

FreeDigitalPhotos.net

Table of Contents

Part Three
Licensing, Patents, Copyrights & Pitching Your Idea

About This Book

IF YOU WERE to peruse the titles in the business section of the bookstore, surf the web or go to your local library, you wouldn't have any trouble locating a large number of informative books that will teach you how to sell or license your idea, your design or your invention....and if you haven't invested in one of those, now is the time to do it. However, the question most frequently posed to me isn't how to sell an idea, but *where* to sell it, a question unfortunately left unanswered in the copious selection of "how to" books.

Akin to a writer's market, Inventor's Market is a directory of manufacturers that buy or license innovative ideas and products from the general population with the intention of bringing them to market. Also included are a handful of product scouts, direct marketing companies*, and informative links to information and services that will benefit the aspiring inventor.

Before contacting a potential partner for your idea, study the company's product lines and their needs to verify that they are a good match for your innovation. Once you have determined the company or companies that have similar interests, make sure that you follow their submission guidelines exactly.

Always verify the integrity of a business and learn the best way to protect your interests before submitting your ideas. For your own protection, it would be wise to consult with an attorney before you sign any contractual agreement.

This book is intended as a directory and at the time of this writing the information is current, but please, be aware, that at some point policies, contact information and website formats may change.

Best of luck on your venture!

*Some of the direct marketing companies require that your product already be manufactured.

The information included on licensing, patents & copyrights is for the reader's benefit, but does not constitute or take the place of professional legal advice.

PART ONE
Manufacturers, Direct Marketers & Product Scouts

Alltrade

Stuff they Sell
Hand Tools, Tool Sets, Mechanic's Tools, Automotive Specialty Tools, Rental / Loaner Tools, Lubrication Tools, Air Compressors and Inflators, Air Tools and Accessories, Power Tools, Power Tool Accessories, Lift / RV / Garage / Shop Equipment, Material Handling / Cargo Management, Power Inverters / Jump Starters / 12V Accessories, Lighting and more.

Website
Home page: http://www.alltradetools.com

Inventor page:
http://www.alltradetools.com/index.php/inventors/

HOW TO REACH THEM

Address
Alltrade Tools LLC
Attn: Senior Legal Counsel
1431 Via Plata
Long Beach CA 90810-1462

Phone Number
(310) 522-9008 (General inquiries)

Email
inventors@alltradetools.com

Some of their products
Cutting & Finishing tools, Measuring tools, Striking tools, Tool storage, Vices and Clamps, Wrenches, Sockets, Steering and Suspension, Engine & Drive-Train, AC corded power tools, Cordless power tools, Bench power tools, Welding & welding accessories, Million candlepower spotlight, and much more.

Some of their brands
Alltrade®, Powerbuilt®, Trades Pro®, Team Mechanix®, Mastergrip®, Air Plus®, Crewline®, Worksmith®, Kawasaki™, HOT ROD®

Idea submission and submission policy
(1) Go to *home page* (2) Go to *inventors* link, or go directly to *inventors* link listed above.

Download Alltrade Tools Idea Submission Policy and Agreement. Sign and mail then notify Alltrade via email that you have mailed in your idea submission agreement.

You will be assigned a project number which must be placed on all correspondence. The next step is to fill out and email Alltrade's Description of Idea or Invention form. The link is located on the inventor's page. Be sure to include your project number.

Comments

If Alltrade Tools has an interest in considering a business relationship with you, it will take about 10 to 12 weeks to review your submission information and get back to you.

Factors taken into consideration when Alltrade reviews ideas and inventions include:

- Is it a carpenter or mechanic's hand tool?
- Is it a power tool, air tool or related accessory?
- Is it automotive lift equipment?
- Is it an automotive specialty tool?
- Does it apply to tool storage and organization?
- If patents have not been awarded, is the invention patentable in and/or outside the United States?
- Does the tool appeal to Do-It-Yourselfers and/or professionals?
- Does the tool make a task easier, faster, safer or more effective?

Aqua Leisure

Stuff they sell
Swim gear and Aquatic leisure products

Website
Home page: http://www.aqualeisure.com/site/

HOW TO REACH THEM

Address
US Corporate Headquarters
Aqua Leisure Industries, Inc
P.O. Box 239
Avon, MA 02322-1098
USA

(Not for submissions)

Phone Number
Toll Free: (866) 807-3998
Fax: (508) 587-5318

(Not for submissions)

Email
customercare@aqualeisure.com

(Not for submissions)

Some of their products
Juvenile fitness equipment, Sun Shade and Protection Gear, Learn to Swim Aids, Swim and Dive Gear, Water Spray Games, and much more.

Some of their brands
First Fitness®, Sun Smart®, Dolfino®, Swim School®, Spray Zone®

Idea submission and submission policy
Aqua-Leisure is always looking for great new product ideas and inventions. If you have one you would like them to consider, go to the website above. At *home page*, go to *Think you have a great idea?* link.

Access their submission form via their link to submit your invention for review.
Please fully complete the step by step online process. If you fail to fully complete the process, and/or fail to consent to the inventor submission agreement contained therein, they will be unable to consider your submission.

Aqua Leisure will acknowledge successful receipt of your invention submission via e-mail. However, please be advised that it may take up to several months for them to further contact you due to the high volume of ideas that they receive and review.

Comments
At the time of this printing, the following products are areas of interest for Aqua-Leisure:

Aquatic fitness equipment and devices
Beach carriers and storage
Beach comfort and convenience
Pool games for interactive and group play
Pool toys
Swim training aids and devices - Adult
Swim training aids and devices - Infant & Juvenile
Water related toys - educational
Water related toys - beach /backyard

These needs may change. To keep updated, go to the submission form link for the most current needs.

Avery Dennison

Stuff they sell
Office products, Decorative labels, Smart (RFID) labels, and Apparel labels

Website
Home page:
http://www.averydennison.com/corporate.nsf/

Innovation page:
http://www.averydennison.com/avy/en_us/Innovation

HOW TO REACH THEM

Address
Idea Submission Program Avery Dennison Corporation
P.O. Box 7090
Pasadena, CA 91109-7090

Phone Number
Does not accept phone calls for idea submissions program
Fax: (626) 792-7312

Some products they make
For the Consumer: Labels, Dividers, Binders & Accessories, Business & Greeting cards, Pens, Pencils & Markers, Sheet protectors,

Notebooks & Paper, Adhesives, Printable craft items, Software.

Business to Business: Automotive, specialty labels, label machines, medical, diaper tapes, architectural, printing, electronics (See website products page for full list and what parts they make.)

Idea submission and submission policy
(1) Go to *home page* (2) Click on *innovations* (3) Click on *Submit an idea.*

Avery-Dennison lists the steps to determine if your idea is suitable for their needs. If you find that your product is a good match, go to the *submit an idea* link on that page, complete the idea submission form and send to the Chief Technology Officer at the above address or fax.

Comments
All communications must be in writing. You will receive a response within 30 days.

Avery Dennison's core competencies include: Adhesives, Web-based Converting, High-Speed Precision, Coating, Printing, RFID, Micro-replication, Extrusion and Films and Functional Coatings.

Ball

Stuff they manufacture
Packaging

Website
Home Page: http://www.ball.com

Submissions Page:
http://www.ballamericas.com/page.jsp?page=3

HOW TO REACH THEM

Address
Listed at site, but N/A for submission

Phone Number
Listed at site, but N/A for submission

Email
inventor@ball.com

Some products they make
Alcoholic beverage bottles that look like glass, metal food and beverage containers, aerosol cans, and more. Ball no longer makes glass products.

Idea submission and submission policy
(1) Go to *home page* (2) Click on *Americas* link (3) Click on *innovation* (4) In side bar, under innovation process, click on *inventors*.

Follow the submission rules described and send your information and request for a submission agreement form via the email link.

Ball will only consider outside ideas, innovations and inventions through the process listed on their idea submission page.

Comments
Ball has different procedures for patented and non-patented submissions. If you have any questions, contact Ball with the email link above. Ball's New Product Steering Committee will review your invention and will then reply in writing if they wish to proceed further.

Current Opportunities at Ball:
Currently ball is developing innovative packaging solutions to help their customers grow in several emerging markets:

Health and Wellness
Ball is developing solutions for smoothies, nutrient-fresh drinks and more. Attention to consumer and brand needs has informed their

design of the Alumi-Tek® aluminum bottle and the slim look of the Sleek™ Can.

Microwave and Convenience

From dashboard dining to convenient home and office meals, Ball is working to expand the market of on-the–go-consumption with products like the Fusion-Tek™ microwavable food can and more.

High-End Beverages

Increased graphic capabilities and innovative options like laser incised tabs provide shelf-differentiation and brand identity that is ideal for high-end beverage markets.

Note: Always check back at the Ball site for updates and changes in what they are looking for.

Bandit Lures

Stuff they sell
Fishing Lures, Caps, Shirts

Website
Home page: http://banditlures.com/

HOW TO REACH THEM

Address
Bandit Lures® Inc.
444 Cold Springs Rd
Sardis, MS 38666

Phone Number
(662) 563-8450

Email
c.ross@banditlures.com

Idea submission and submission policies
Bandit Lures will look at and consider new products. Depending on originality, design and whether it can be patented, they will pay royalties. They will not pay for obvious knock-off's/copies of other products.

Comments
If you have a good product, call Chris

Ross/CEO at the above phone number.

Bard

Stuff they sell
Health Care Products in the fields of: Vascular, Urology, Oncology, and Surgical Specialty.

Website
Home page: http://www.crbard.com/

Submission page:
http://www.crbard.com/products/ideageneration brochure/index.cfm

HOW TO REACH THEM

Address
Vice President of Science and Technology
C.R. Bard, Inc.
730 Central Ave.
Murray Hill, NJ 07974

Phone Number
(908) 277-8185
(908) 277-8000
Fax (908) 277-8363

Email
john.deford@crbard.com

Some products they make
Ablation catheters, Bone biopsy products,

Carotid shunts, Angioplasty balloons, Collagen implant, Cutting loops, Epidural catheters, Core tissue biopsy products, Guide wires, Hem dialysis catheters, Patient monitoring systems, Pouches, Radioactive seed implants, Stone baskets, Skin and wound care products, and much more. See website alphabetical listing of products for complete list.

Idea submission and submission policy
(1) Go to *home page* (2) In left column click on *submit a product idea online* (3) This will take you to Idea Generation Process page with links to Bard's Idea Generation Brochure and online idea submission form. Fill out submission form and submit to Bard. Bard will respond within 30 days.

Comments
Submission can be made online or sent through the mail.

Bayer Animal Health

Stuff they sell
Products for Animal Health

Website
Home page:
http://www.animalhealth.bayerhealthcare.com/

Submission page:
http://www.animalhealth.bayerhealthcare.com/3446.0.html

HOW TO REACH THEM

Address
N/A

Phone Number
N/A

Email
Click on *Contact us* and email comments and questions

Some of their products
Anti-infective, Disinfectants, Pharmacological, Insecticides, Rodenticides, Endectocides.
(See website for complete list.)

Some of their brands/trademarks
Advantage, Advantage Multi, Advantix, Advocate, Baytril, Baycox, Droncit, Tempo, Solfac, Vircon. (See website for complete list.)

Idea submission and submission policy
Go to submission page listed above or (1) Go to *home page* (2) Click on *contact license* listed under *Business Development* link then fill out and submit form with product invention information.

Comments
This is a licensing opportunity.

Note: If you disclose information to Bayer which is not already the subject of a filed patent application, your disclosure may be considered to be a publication and may jeopardize any further patent application.

BD

Stuff they sell
Medical supplies, devices, laboratory equipment and diagnostic products

Website
Home page: http://bd.com/

Submission info page:
http://bd.com/support/contact/newideas.asp

HOW TO REACH THEM

Address
Office of New Product Suggestions
BD
21 Davis Drive
P.O. Box 12016
Research Triangle Park, NC 27709-2016

Phone Number
N/A for submissions

Email
N/A for submissions

Some of their products
Microbiology products: Adapters and tubing, Blood culture, Collection & transport, and more.

Injection products: Alcohol swabs, sharps collector, biopsy needles, pharmacy needles, syringes, and more.

Diabetes care products: Insulin syringes, insulin pens, lancets, and more.

Surgical products: Specialty blades, scalpels, antimicrobial soaps and scrubs.

This is only a fraction of BD's product categories and products. To see them all, go to their product catalog link on their website.

Idea submission and submission policy
At home page, (1) go to *Contact BD* link (2) Click on *New Product Ideas,* or go directly to New Product Submission page listed above.

Following the guidelines, fill out and sign the form provided in their PDF document, *Consideration of Submissions,* provided via a link on the New Product Submission page. Make a duplicate for your records. Then send the completed form to the company along with a disclosure of your suggestion.

The disclosure should be as complete as possible so the company will have adequate information to decide whether or not it is interested.

If possible, publications, published copies of the patent or patent application including claims and drawing would be helpful.

A copy of the disclosure should be kept by the submitter since it is necessary that they keep the submitted material for reference in case a question should arise as to what was disclosed.

BD cannot accept any prototypes; only drawings and/or literature will be accepted. All mailings must be in flat envelopes no bigger than 11x14 inches and any packages or mailings not fitting this description may be destroyed without review.

Comments
BD accepts innovative ideas for new products and technologies to their company by individuals and organizations not employed by or affiliated with BD. Inventions must be covered by existing patents or have a patent application submitted.

BD presently cannot accept e-mail submissions. If you are unable to print the document, please write to the Office of New Product Suggestions (above) and request a copy

Decisions take 4-6 weeks.

Question:

What is an NDA or Non-Disclosure Agreement?

Answer:

A Non-Disclosure Agreement, also referred to as a NDA, is an agreement between two or more parties to protect and keep confidential the information that is covered in the NDA. For example, if you submit a product or invention idea to a manufacturer, an NDA will protect your idea from being shared with parties outside of those listed in the agreement without your authorization.

You can provide your own non-disclosure form if the manufacturer is agreeable to signing it. However in most cases, manufacturers have their own confidentiality agreements that require your signature before an idea can be submitted.

Bel-Art

Stuff they sell
Science products/scientific lab equipment

Website
Home page:
http://www.belart.com/shop/index.php

Submission page:
http://www.belart.com/shop/inventor.php

HOW TO REACH THEM

Address
Bel-Art Products
6 Industrial Road
Pequannock, NJ 07440

Phone Number
(800) 4BEL-ART
(973) 694-0500
Fax (973) 694-7199

Email
feedback@belart.com

Some products they make
* Note: BelArt's products are for scientific labs.
Biohazard bags, Disposable bags, Beakers, Bottles, Centrifuge ware, Clamps and Holders, Flasks, General laboratory, Glassware

accessories, Pitchers, Labeling, Racks, Stoppers, Trays, Glove boxes, Pipetting, Siphons and Pumps, Funnels and Filters, and more.

Idea submission and submission policy
Click on the above submission information page or (1) Go to *home page* (2) Click on *inventors corner* in the left column.

Provide Bel-Art with a few general statements about your idea or invention. Please include information like end user market and why you feel your product would be well received in the marketplace.

Contact them at feedback@belart.com or call 1-800-4BEL-ART

Comments
Products are introduced and sold in Scienceware® Tools for Science Catalogue.

Your innovation can be new or can be an improvement on an existing invention. Bel-Art says to contact them if you believe that your idea or invention would benefit the science community; if you believe that your idea or invention would be marketable to the science community; have an interest in collaborating with experienced professionals, or would simply like to explore the possibilities.

Bench Dog Tools

Stuff they Sell
Woodworking Tools

Website
Home Page:
http://www.benchdog.com/index.cfm

Inventor Page:
http://www.benchdog.com/inventors.cfm

HOW TO REACH THEM

Address
Bench Dog Tools
9775 85th Avenue North
Maple Grove, MN 55369

Phone
(800) 786-8902
Fax (763) 515-7463

Email
info@benchdog.com

Some of their Products
Cast Iron ProMax RT, Phenolic Pro Top Router Tables, ProStand, ProLift, ProCabinet, Router Table & Table Saw, Feather Loc® Multipurpose Featherboard, and much more.

See website product catalog.

Idea submission and submission policy
Go to *Home Page* and click on *Inventor link*, or go to *Inventor Page* above for more information on submissions.

You can contact Bench Dog Tools at the address, email or phone number included on this page. Bench Dog will talk about your general idea, and determine whether or not they already have something similar in development. If they do not, they will enter a confidentiality agreement.

Comments
Bench Dog recommends you consult with an attorney before discussing your ideas. The best time to submit ideas is during the summer.

Benjamin Obdyke

Stuff they sell
Construction industry/Sidewall products, Roof products

Website
Home page: http://www.benjaminobdyke.com/

Submission page:
http://www.benjaminobdyke.com/content/submitIdea

HOW TO REACH THEM

Address
Benjamin Obdyke Incorporated
400 Babylon Rd., Ste A
Horsham, PA 19044

Email
info@benjaminobdyke.com

Phone Number
(215) 672-7200

Some of their brands/products
Home Slicker®, Home Slicker® Plus Typar®, Mortairvent®, Plypatch®, Waterfall Gutter guard®, Cedar Breather®, Xtactor Vent®,

Rapid Ridge®, SynShield®, Roll Vent®, and more

Idea submission and submission policy
Go to submission link above or (1) Go to *home page* (2) Go to *Have an Idea* link (3) Under *Idea Submission* go to *what is benovation* for complete details on product idea submission.

Comments
Benjamin Obdyke is looking for product and service ideas. If your idea is already patented, you can bypass the confidentiality form and begin by submitting the idea submission form.

Contact information for Confidentiality Agreement:

Attention Director of product development:
Phone (215) 672-7200

Big Idea Group

What they do
Big Idea Group (BIG) connects creative inventors with innovation-seeking companies. They represent inventors, taking on the role of inventor's agent or they will opt to license new product ideas.

Stuff they represent
BIG has an interest in innovative consumer product goods, but they have a particular interest in hardware, infomercial, promotional, house wares, pets, sporting goods, and lawn & garden.

Website
Home page: http://www.bigideagroup.net/

HOW TO REACH THEM

Address
Big Idea Group
175 Canal Street, 5B
Manchester, NH 03101

Phone Number
Fax: (603) 641-5995

Email
Info@BigIdeaGroup.net (For general inquiries)

Idea submission and submission policy
View sample submissions at the following link: http://www.bigideagroup.net/inventors/sample_submission.htm

You will need to fill out a pre-submission registration form and entry agreement before you submit your idea. Follow the instructions and then mail or fax your invention information. Go to inventor link for complete details.

Comments
You can present your invention/product at road shows (See website for dates and locations), phone-in events, or general mail submission.

Big Idea welcomes ideas in all stages of development . Once you have pitched your idea they will contact you with a decision within 45 days.

Big Idea is not interested in videos, board games, books, music, CDs, and licensed products. Inventor contests are also frequently found on Big Idea's website.

Black & Decker

Stuff they sell
Power Tools, Automotive, Portable Power, Cleaning Products, Painting and Craft, Appliances

Website
Home page: http://www.blackanddecker.com/

Submission page:
http://www.blackanddecker.com/CustomerCenter/Product-Ideas.aspx

HOW TO REACH THEM

Address
Inventions
The Black & Decker Corporation
701 E. Joppa Road
Towson, MD 21286

Phone Number
FAX (410) 716-2788

Email
N/A

Some of their products
Drills, Grinders, Routers, Rotary Tools, Sanders, Screwdrivers, Laser levels, Stud sensors and detectors, Measuring tools, Saws,

ols, Blower vacs and Sweepers, Chain s, Scrubbers, Lawn mowers, Pressure washers, Gardening tools, Batteries and Chargers, Canister vacuums, Stick vacuums, Automobile vacuums, Cordless vacuums, Hard floor cleaner, Glue guns, Power scissors, Paint application tools, Weather radio, Battery chargers, Power inverters, and much more. See website product guide.

Idea submission and submission policy
Submissions from the US and Canada may only be submitted online. Go to submission page link above or (1) Go to *home page* (2) Go to *product ideas* at bottom of the page. This link will take you to the submission page where you can propose your idea via their submission system and will also provide a link to B & D's idea submission brochure. This brochure offers more details on what their expectations are.

Comments
Your idea needs to be unique, have commercial viability and fit within Black & Decker. The Auto wrench, Auto tape, and Alligator are just a few ideas that Black & Decker has picked up through inventor idea submissions.

Presently, Black & Decker is not accepting ideas on inventions related to garment care or food preparation appliances.

If you are not from the US or Canada, go to the external submissions link on the submissions page for Black & Decker's submission process.

Black & Decker will not evaluate any ideas or inventions submitted via the external submissions process if it is not protected by a patent or patent application.

Bobcat

Stuff they sell
Compact equipment for construction, Agriculture, Rental, Landscape, Grounds maintenance, Mining, Industrial

Website
Home page: http://bobcat.com/

HOW TO REACH THEM

Address
Bobcat Company
Attn: Attachments Department
P. O. Box 6000
West Fargo, ND 58078-6000

(Submission address)

Phone Number
(701) 241-8700

Email
N/A for product submissions

Some of their products
Attachments, Parts & Services, Loaders, Excavators, Utility Products, VersaHandler® TTC,

Idea submission and submission policy
Bobcat is always interested in Bobcat attachment ideas. They do require the signing of an Invention Submission Form before they will accept the disclosure of a new invention.

To access Invention Submission Form link, go to *Home Page* and click on the *Questions* link,

...or at *Home Page*, under *Company Information,* go to, *Contact Us*.

Scroll down to: *I have a bobcat attachment idea*...

...or go directly to http://bobcat.com/our_company/contact_us/faq and do the same.

If you do not have a patent, or an application for a patent has not been filed, you should have your drawings signed, dated and preferably witnessed.

The submission form states that a full written disclosure must be furnished to the company, preferably the patent application drawing and specification, if one exists, or if not, a rough sketch or drawing provided it illustrates the invention so one skilled in the art can understand it.

Bobcat recommends you keep a duplicate of what you send them.

Comments

Things Bobcat wants you to consider before submitting your idea:

- Does the attachment or product idea have patent protection?
- Do you want to sell the rights to the attachment, manufacture it yourself, or license the rights to the attachment with a royalty?
- Have any market studies been done for such an attachment?
- Has a patent search for similar attachments or ideas been done?
- What type of development & test time has the product had?
- After you have reviewed the checklist, please complete and sign the Invention Submission form. Send it along with a full written disclosure of the attachment and any available literature, pictures and video as well as a copy of any patent to the address listed above.

If you wish to contact Bobcat by phone to discuss your attachment idea, call: (701) 241-8700 and ask to speak with Phil Bogner, but please complete the steps listed above ahead of time.

Bosch

Stuff they sell
Car parts, communications, household appliances, garden tools, repair shop diagnostics, car parts, power tools, automotive technology, heating and warm water, security systems, power tools for professionals, packaging technology, sensors and foundry-MEMS

Website
Home page: http://www.bosch.com/

Submission/purchasing logistics page: http://purchasing.bosch.com/en/start/Einkauf/innovation/index.htm

HOW TO REACH THEM

Address
Robert Bosch GMBH Postfach 106050 70049
Stuttgart
Germany

Correspondence for an offer should be addressed to: Robert Bosch GMBH
Corporate Intellectual Property Licensing
P.O. Box 30 02 20
D-70442 Stuttgart
Federal Republic of Germany

Phone Number
+49 (0)711 811-0

Some of their products
Car batteries, Starters, Wiper blades, Air filters, Brake systems, Lawn mowers, Pressure cleaners, Freezers, Dishwashers, Washers, Dryers, Ovens, Ranges. Bosch's product list is extensive and can be searched out online.

Idea submission and submission policy
Bosch is interested in innovations at various stages of development. For products already in production or that have only been developed in your mind or produced via CAD, (1) Go to *home page* (2) Go to *purchasing and logistics* link (3) Go to *purchase your innovation*. (http://purchasing.bosch.com/en/start/Einkauf/innovation/index.htm)

Bosch requests that you:

- Describe your proposal as short and precise as possible.
- Give a keyword for how far the development is already advanced: model, prototype or already produced in series.
- Tell them if and how your idea is protected by patent rights.
- Convince them with the realistic evaluation of the economic importance of your idea.

* Mention strengths and weaknesses, chances and risks, unique selling points and competitors.

  You will find guidelines for technical innovation providers via the *Bosch Your invention partner* link.

To submit your idea, go to *online formular* link and fill out required form.

If you have a question or want to submit your patent, your contact is: Silke Scheitler, Licenses

Email: Silke.Scheitler@de.Bosch.com

Comments
Bosch only considers ideas that are offered in sufficient technical detail and are the subject for a patent application. In exceptional cases they will review ideas that are not the subject of a patent application if the inventor bears the risk.

This website is offered in German and English, if you visit a page and it is not in your language, there is a link located at the top where you can change it.

The Bradford Group

Stuff they sell
Collectible Brands

Website
Home page: http://www.thebradfordgroup.com

HOW TO REACH THEM

Address
The Bradford Exchange
Licensing and Artist Relations
9333 Milwaukee Avenue
Niles, IL 60714

Phone number
(847) 966-2770

Email
artinquiry@bgeltd.com

Some of their products
Ornaments, Miniature plates, Teacups, Sculptures, Porcelain dolls, Vintage dolls, Bride dolls, Collector plates, Music boxes, Sports, Collectibles, Die cast cars and trucks, Handcrafted architectural and cottage miniatures, Collectible electric trains, Curio cabinets, Ornament displays, Figurines.

Some of their brands
The Bradford Exchange, Ltd., The Ashton Drake Galleries, Ltd., The Bradford Editions, Bradford Authenticated, Ardleigh Elliott, The Hamilton Collection, Van Hygan and Smythe, Hawthorne Village, Gallery Marketing Group, Hamilton Authenticated, Collectibles Today.

Product submission and submission policy
For consideration by The Bradford Group's Art Review Committee, send one of the following as reference to your work: digital documents of your artwork on a CD, printed color references, or a duplicate set of ten to twenty 35mm slides. Please include a resume and any additional information about your work and experience. Please do not send original artwork or items that are one of a kind.

Samples of your work may be kept in their Artist Resource File and reviewed by Product Development Team members as product concepts are created that would be appropriate for your work.

Comments
The purpose of this listing is for licensing your artwork, however, The Bradford Group also hires artists for contract work, and of course for employment.

Email: careers@thebradfordgroup.com

Cactus Marketing

What they do
International toy and game inventor agents

Website
http://www.cactusmarketing.com/

HOW TO REACH THEM

US Information
Address
1553 South Military Highway
Chesapeake, VA 23320

Phone Number
+01 (757) 366- 9907
Fax +01 (757) 366-9913
Toll Free (U.S.A. only) (888) 215-7040

Email
cactuspete@cavtel.net

UK Information
Address
109A Hamilton Road
Felixstowe
Suffolk
IP11 7BL
England

Phone Number
+44 (0)1394 275275
Freephone (U.K. only) 0(800) 389-8408

Email
games@cactusmarketing.com

Some of their licensed products
Scattergories™, Dyrenes Hotel, Super Scattergories™, Rainbow Timber,

Idea submission and submission policy
Go to *submit your idea* link. To submit your idea for consideration, contact Cactus Marketing using the contact information for your area to receive an information package on the submission procedure.

Do not send your idea without first receiving their submission guidelines and details of terms and conditions which are included within the information package.

Comments
Cactus is able to present your item to key decision makers of major toy and game companies world-wide for licensing consideration.

As well as attending the major international toy fairs in London, Nuremberg and New York they make numerous individual presentations promoting their client's toy and game inventions.

Carlon/
Lamson & Sessions

Stuff they Sell

Fiber optic, copper and coaxial cable protection for plant construction and wiring systems, (LHP) sells to the do-it-yourself, hardware and mass merchandiser markets, providing homeowners and contractors with rough electrical, convenience and security products.

Website

Home Page: http://www.carlon.com

HOW TO REACH THEM

Address

Lamson & Sessions
8155 T&B Blvd
4B-36
Memphis, TN 38125

Phone Number

(800) 816-7809 (Customer Service)
Fax: (800) 816-7810 (Customer Service)

Email

www.carlon.com

Some of their Products
Wall Sconce, Mini Spotlights, Chimes, Switch and Outlet Boxes, Home Security Alarms, Holiday Items, and more.

Idea submission and submission policy
From *Home Page*, go to *New Product Ideas* link and download Policy and Agreement Concerning Ideas Submitted by Persons Outside of the Lamson and Sessions Company form. This must be sent along with your idea to the above address.

Celebrating Home

Stuff they sell
Wall Décor, Home Accents, Home Fragrance, Dining & Entertaining, Patio & Garden, Decorator Florals, Gourmet Food.

Website
Home page: http://www.celebratinghome.com

HOW TO REACH THEM

Address
Celebrating Home
2938 Brown Road
Marshall, TX 75672

Phone Number
(800) 700-7873

Email
Sfloyd@celebratinghome.com
customerservice@CelebratingHome.com
designercare@celebratinghome.com

Some of their products
Dips, Desserts, Jams, Preserves, Decorator floral arrangements, Veranda Serving Bowls, Veranda Heart Baker, Casseroles, Fondue set, Canister sets, Pitchers, Party Servers, Dinner plates, Bake ware, Serving Pieces, Flameless

Candles, Jar Candles, Pillar Candles, Candle Holders and Accessories, Figurines, Clocks, Furniture, Bath accessories, Sconces and Metal Wall Décor, Framed Art, and much more. See Website.

Idea submission and submission policy
Please submit any new product ideas you have to Susan Floyd in Celebrating Home's Products department. She can be contacted at: Sfloyd@celebratinghome.com

Clorox

Stuff they sell
Laundry Products, Household Cleaning Products, Barbeque Sauces, Charcoal Brands, Car Care Products, Salad Dressing, and Water Filtration Pitchers

Website
http://www.clorox.com/

HOW TO REACH THEM

Address
The Clorox Company
P.O. Box 24305
Oakland, CA 94623-1305

Phone Number
N/A

Email
Website email form:
http://www.clorox.com/contact.php

Some of their products
Clorox® regular bleach, ToiletWand System™, Disinfecting Wipes, Anti-allergen bleach, Outdoor bleach cleaner, Ready Mop ® Mopping Systems, Disinfectant Spray, Disinfectant kitchen cleaner, BathWand

System, Blue automatic toilet bowl cleaner, Fresh Care™ towels, ProResults bleach, and more. See website for complete list.

Some of their brands
Clorox® Oxi Magic™, Green Works™

Idea submission and submission policy
Contact Clorox at their webpage email and request information for idea submissions. You should receive their Express Disclaimer of Confidential Relationship Form, along with an informative cover letter. Fill out form and mail. This must be received by Clorox before they will review your idea.

The Coca-Cola Company

Stuff they sell
Beverages

Website
Contact us page:
http://www.thecoca-colacompany.com/contactus/

Submission link page:
https://secure.thecoca-colacompany.com/ssldocs/contactus/cokesubmit/cokesubmit.shtml

HOW TO REACH THEM:

Address
The Coca-Cola Company
P.O. Box 1734
Atlanta, GA 30301
USA

Phone Number
N/A

Email
N/A

Some of their products
Energy drinks, tea, coffee, bottled water, sports drinks, soft drinks, juices and juice drinks.

Some of their brands
A&W, Bright & Early, Crush, Big Tai, Coca-Cola, Crystal, BlackFire, Cresta, Bistra, Club, Calypso, Dasani, Fruitia, Earth & Sky, Five Alive, Dr. Pepper, Diva, Eva Water, Inca Kola, Hi Spot, Hawaii, Fresca, Fanta, Godiva Belgian Blends, Ice Dew, Minute Maid Splash, Morning Deli, Minute Maid, Mr. Pibb, Powerplay, Powerade alive, Just Juice, Sobo, Swerve, Squirt. The Coca-Cola brand list is extensive. To view them all go to website above then link to *brands,* and again link to *brand list*.

Idea submission and submission policy
Go to the *contact us* page and click on the, *submit an idea* link. This will take you to terms and conditions and proper online submission forms.

Comments
Coca-Cola prefers that all submissions be protected by a patent, copyright or trademark, depending on product idea type.

Coca-Cola must abide by their policy that prevents consideration of ideas related to advertising, formula modifications to any of their existing products and recurrent concepts they have previously considered.

Coleman

Stuff they sell
Outdoor Sports/Camping Equipment

Website
Home page: http://www.coleman.com/

HOW TO REACH THEM

Address
The Coleman Co., Inc.
Attn: Legal Dept. - Invention Evaluation
3600 N. Hydraulic
Wichita, KS 67219

Phone Number
(316) 832-2653

Email
N/A

Some of their products
Grills, heaters, insect repellants, sleeping bags, Fishing, Boating, Tail-gating, Hunting, Tents, Storage, Water Sports, ATV accessories, Coolers, Watches, Stoves, Heaters, Back Packs, Screened canopies and more.

Some of their brands
Hodgman, Sevylor, Stearns

Idea submission and submission policy
In order to protect inventors and Coleman, only patented inventions will be considered. You may send Coleman a copy of your issued patent for review to the above address. Please do not send any confidential information with your patent. We will not review any patents pending.

Further information about Coleman's policy can be obtained by calling (316) 832-2653. Once the recorded message begins, please select Option 6, followed by Option 5.

Creative Teaching Press

Stuff they sell
Educational books and products for grades Pre-K-8th
(Submitters must have an educational background)

Website
Home page: http://www.creativeteaching.com/

Submission page:
http://www.creativeteaching.com/t-SubmitAnIdea.aspx

HOW TO REACH THEM

Address
Forward all submissions to:
Creative Teaching Press
Attn: Idea Submissions
15342 Graham Street
Huntington Beach, CA 92649

Phone Number
(800) 287-8879
(714) 895-5047

Email
Emails can be sent via their contact us page

Some of their products
Workbooks: Math, grammar, reading, etc.
Teacher resource books: Art, Language arts, Health and self esteem, Character education.
Learning Décor: Banners, borders, Cut outs, etc. See website product catalog for complete list.

Idea submission and submission policy
If you have a proposal or manuscript you would like to be reviewed, please submit the following:

In a cover letter, prepare a brief description of the material you want to have considered for publication, including grade level and a synopsis of your background as an educator. Your idea is more likely to be considered if you have identified a specific need in the educational marketplace and an explanation of how it is unique to the competition.

Submit your manuscript or product idea and include a summary of your material, table of contents and at least one chapter. If you have an idea that is not yet fully developed they will review an outline and a representative sampling of your work. However, material that is fully developed and classroom tested is more likely to capture attention.

Please include a self addressed stamped

envelope with sufficient postage to return ideas if they do not fit Creative Teaching Press' publishing plans.

Comments

What they are looking for:

Creative Teaching Press publishes a wide variety of products for grades PreK–8, including teacher resource books, bulletin boards, borders, emergent readers, and charts.

They are always interested in reviewing new ideas developed by teachers and are especially interested in book proposal submissions that include ideas that have been successfully tested and used in the classroom.

Currently their greatest area of interest is in the PreK–4 grade range.

Crown

Brand Building Packaging™

Stuff they sell

Packaging for a variety of product areas including, Food, Beverage, Health & Beauty, Industrial & Household

Website

Home Page: http://www.crowncork.com/

Submission Page: http://www.crowncork.com/innovation/innovation.php

HOW TO REACH THEM

Address

Corporate Address:
Crown Holdings, Inc.
One Crown Way
Philadelphia, PA
19154-4599 USA

Phone Number

(215) 698-5100

Email

Page for email submission:
http://www.crowncork.com/innovation/info.php

If you have questions about submissions or are following up on a submission you have made, please go to *More Information* link or use the above site page for your inquiries.

Some of their products
Aerosol Containers, High Impact Decoration & Finishes, Shaped Containers, Specialty Containers, Metal Closures, Food Containers, Crowns, Beverage Cans, and more.

What they are looking for
Crown is looking for a variety of technologies that will improve shelf life, improve manufacturing, provide branding opportunities, pack enhancement for consumers, new packaging technologies, improved sustainability in several areas, including, but not limited to:

Animated Graphics, Aerosol Atomization and Delivery, Low Energy Curing of Coatings, Corrosion inhibitors, sealants (materials and application methods), materials and processes for heat sealing, technologies for improved canned food, freshness indicators, technology for self heating and self cooling cans, RFID, and much more.

For complete information, please go to *Areas of Interest* link, or go directly to areas of interest page at:

http://www.crowncork.com/innovation/interest.php

Idea submission and submission policy
(1) From home page, go to *innovation & design* link or go directly to submission link listed above (2) Go to *open innovation* link.

Review Crown's *Areas of Interest* page. Read and acknowledge the submission terms. Complete the Submission Form and upload any necessary non-confidential supporting information. You will get an email confirmation that they have received your inquiry.

Assessment: Crown's Technology and Innovation teams will review the Submission Form and supporting materials and determine their level of interest.

Response: Crown's goal is to evaluate the submission and respond to you within 30 days of receiving your material however the process may take longer.

Comments
If you have an idea that is not listed, but you believe it would fit the needs of Crown, Crown

encourages you to share it with them.

Their policy is to accept only non-confidential information at the initial stage. Should Crown decide to pursue your submission, they may require you to sign a confidentiality agreement before there is further exchange of information.

Dewalt

Stuff they sell
Power Tools, Attachments, Accessories, Metal Working, Laser and Instruments, Pressure Washer Accessories, and more.

Website
Home page: http://dewalt.com/Home.aspx

Submission page: http://dewalt.com/Company-Information.aspx

HOW TO REACH THEM

Address
Inventions
DEWALT Industrial Tool Co
701 E. Joppa Road, MS TW075
Baltimore, MD 21286

Phone Number
1-800-4-DEWALT

Email
N/A

Some of their products
Drills, Hammer drills, Portable table saws, Grinders, Tool bags, Tool saw attachments Laser attachments, Saw blades, Lighting,

Metalworking, Drywall bits, Pump oils, Spray guns, Water brooms, Extension wands. To see full list, visit their website and go to products at a glance.

Idea submission and submission policy
Go to submission page link above, or (1) Go to *home page* (2) Go to *company info* link at bottom of page (3) Click on *invention submission* link.

This link will take you to the online Dewalt proposal submission form. There will also be a link to Dewalt's idea submission brochure. This brochure offers more details on what their expectations are.

Comments
Your idea needs to be unique, have commercial viability and fit within Dewalt.

If you are not from the US or Canada, go to the external submissions link on the submissions page for Dewalt's submission process.

Dewalt will not evaluate any ideas or inventions submitted via the external submissions process if it is not protected by a patent or patent application.

DexBaby

Stuff they Sell
Child Safety and Comfort Products

Website
Home Page: http://www.dexproducts.com/

Submission Page:
http://www.dexproducts.com/ideas.htm

HOW TO REACH THEM

Address
DEX Products, Inc.
840 Eubanks Drive, Suite A
Vacaville, CA 95688 USA

Phone Number
(800) 546-1996
(707) 451-7864
Fax: (800) 546-1057
(707) 451-8758

Email
mail@dexproducts.com

Some of their products
EZ™Bather, Folding Changing Pad, Wipe Warmer Deluxe, Safe Sleeper ™ Bed Rail, Baby Food Processor, Automobile Bottle

Warmer, Safe Lift, Pregnancy Pillow, Secure Sleeper™, Universal Safety Gate, Space Saver™ Wipe Warmer, and much more. See site for complete list.

Idea submission and submission policy
At *Home Page* go to *submit ideas* at bottom of the page or go directly to submission link above.

Contact DexBaby by phone, email, or mail to find out how they want you to submit your information. You must contact them before submitting your product idea.

Comments
DexBaby is always on the lookout for new ideas to license.

Erico

Stuff they sell
Electric **R**ailway **I**mprovement **CO**mpany (ERICO®) Precision engineered specialty metal products for electrical, commercial & industrial construction, utility & rail applications.

Website
Home page: http://www.erico.com/

Submissions page:
http://www.erico.com/static.asp?id=15

HOW TO REACH THEM

Address
Send submissions to:
Erico, Inc.
Engineering, Idea Submissions
34600 Solon Rd.
Solon, OH 44139

Phone Number
(440) 248-0100
Fax (440) 248-0723

Email
N/A

Some of their products
Electrical fixings, fasteners and supports: Drywall, hangers, Ceiling/partitions, Acoustical, stud wall, Miscellaneous components.

Rail and industrial products: Cable clips, Grinders and equipment, Grounding, Mechanical connecters, Rail drilling machines and much more. For full list see Erico's products page.

Some of their brands
Caddy®, Eritech®, Eriflex®, Lenton®, Cadweld®, Critec®

Idea submission and submission policy
Click on submission page link listed above or (1) Go to *home page* (2) *Contact us* (3) Click on *idea submission* link (4) Read directions and if applicable, go to *idea submission forms,* fill out and submit.

Comments
Erico prefers to have the idea or device protected before it is submitted. They will not consider ideas that cannot be patented.

Erico will consider ideas that relate to new or improved products, manufacturing methods or machinery. These are ideas that if appropriate for them could be patented.

Eureka Medical

What they do
Eureka Medical is an Inventor Network for medical inventors. They offer no cost services to help refine and present medical product ideas to the best matched medical supply companies.

Website
Home Page: http://www.eurekamed.com/

HOW TO REACH THEM

Address
Eureka Medical, Inc.
3434 East Bengal Blvd. Suite 328
Salt Lake City, UT. 84121

Phone Number
(781) 229-5878 (Donna Voiland)
Fax (617) 812-0094

Email
invent@eurekamed.com

Products they are looking for
Anything medically related

Idea submission and submission policy
Eureka Medical Invention Reviews are free and conducted throughout the United States. They will arrange a free, private, confidential session with a panel of medical device experts, who will evaluate your medical innovation. If you can't attend in person, they will schedule a phone conference or evaluate your idea by mail.

You'll receive constructive advice at no charge along with the possibility that Eureka will represent your medical invention to innovation-seeking medical supply or medical equipment distributor companies.

Register one of four ways.

(1) Send your Invention Executive Summary and signed Agreement to the above address. These forms can be found and printed out on the website under *Inventor Legal Agreements.*

(2) Fax it to the listed fax number

(3) Call the client relations manager at the above number or…

(4) email the following information:

Name, contact Information, road-show location & date you are interested in attending. (Road show schedule is on home page.) Brief summary of your professional background, name of your invention, area of focus or specialty, distinguishing achievements such as number of patents and commercialized inventions

Comments
Eureka Medical recommends you seek legal counsel before signing their contract.

Exceptional Products, Inc

What they do
Exceptional Products is Direct Response Television

Stuff they market
Automotive, Beauty and Personal Care, Educational and Self Improvement, Entertainment and Travel, Health and Fitness, House Wares and Electronics, Sports, Hobbies, Collectibles, Weight Loss/Diet and Business Opportunities.

Website
Home page: www.sellontv.com

Submission page:
http://www.sellontv.com/about_us.aspx

HOW TO REACH THEM

Address
Exceptional Products, Inc.
12250 Inwood Road, Suite 6
Dallas, TX 75244

Phone Number
(800) 536-5327
(972) 387-8077 (Outside U.S.)

Fax: (972)387-0515
Contact: George Asmus, phone ext. 104

Email
newideas@sellontv.com

Some of their Products
Hairdini®, Wrapsody®, Wrap/snap/go®, Save a Blade™, Teen hairdini®, All Gone Stain Remover, Pillow Rest.

Idea submission and submission policy
Go to submission page link above or (1) Go to *home page* (2) Click on *about EPI* (3) Read about their procedure and click on *submit* link to submit idea via their web page.

Call or email for further information.

Comments
This is direct response television (DRTV), where accepted products are marketed on TV.

EPI states there are no upfront charges or fees and that they make money only when your product sells.

Faultless

Stuff they sell
Faultless seeks new products to distribute to the mass market in the following areas: Cleaning, Garment Care, Gifts, Hardware, Home Improvement, House Wares, Automotive, Outdoor Living and Lawn and Garden consumer products.

Most consumer product categories are accepted by Faultless, excluding apparel, industrial, food & drug or other products with considerable environmental or product liability risks.

Website
Home page: http://www.faultlessinventors.com/

Submission page information:
http://www.faultlessinventors.com/submit_invention.asp

HOW TO REACH THEM

Address
Faultless Inventors
1025 W. 8th Street
Kansas City, MO 64101

Phone Number
N/A

Email
N/A

Some of their products
Starch, Wrinkle remover, Hot iron cleaner, Stainless steel and Copper cleaner, Aluminum cleaner, Polishing cleaner, Cleaning powder, Candles, Room sprays, and much more.

Some of their brands
Magic®, Bon Ami®, Steel Glo®, EZ Off™, Faultless Starch®, Handybar®, Kleen King®

Idea submission and Submission policy
Go to Faultless submission webpage or (1) go to *home page* (2) go to, *submit your product* link.

Choose the appropriate link for where your product idea is at pertaining to patents or being market ready and fill out the submission form.

Comments
Faultless sells products through the mass market stores, independent retailers, the internet, catalogs and television. They offer product development assistance, design and sourcing, their retail buyer network and new product marketing campaigns. Faultless

purchases your product or will license your invention and pay you a royalty.

Faultless is presently seeking inventions that are patented, patent pending, or have provisional patent protection.

Faultless also sponsors an annual invention contest. The link for contest information can be found on their website's *home page*.

Fellowes

Stuff they sell
Business Machines, Office Productivity, Record Storage, Computer Accessories

Website
Home page: http://www.fellowes.com/

HOW TO REACH THEM

Address
Fellowes, Inc
Product Idea Attn: Karen Aiello
1789 Norwood Ave
Itasca, IL 60143

Phone Number
(630) 893-1600 (General questions)

Email
N/A for product submission

Some of their products
Shredders, Cutters, Keyboard trays and drawers, Wrist supports, Monitor Supports, Copy Holders, Back Supports, Foot Supports, Desk Organizers, Literature Organizers, Cubicle Organizers, Mouse-pads, Keyboard Guards, Rotary Trimmers, Storage boxes and more.

Idea submission and submission policy
Please send a written proposal for your product idea by mail to the address listed above.

Fiskars

Stuff they sell
Tools for School, Office, Crafts and Gardening

Website
Home page: http://www.fiskars.com

HOW TO REACH THEM

Address
Fiskars
ATTN: Product Development
2537 Daniels Street
Madison, WI 53718

Phone Number
(866) 348-5661

Email
Not available for product submission

Some of their products
Adhesive foam strips, Glue sticks, Cutting mat, Bow maker, Utility knife, Craft hand drill, Wired pliers, Card crafting kits, Sharpeners, Paper crimpers, Decorative scissors, Shape cutters, Craft caddy, Trimmers, Stamping, Scissors, Project totes, and much more. See website for full product line.

Idea submission and submission policy
Send your product idea to product development department above.

For further inquiries, submit a general inquiry form under the contact information page, http://www2.fiskars.com/Customer-Service/General-Inquiry

Comments
Your product idea must have a patent.

Flambeau

Stuff they sell

Flambeau manufactures a variety of products through several lines.

Arts and Crafts products, Toys, Hardware storage products, Fluid systems, Hunting & Fishing products, Lawn ornaments, Medical and industrial storage containers

Website

Home page: www.flambeau.com/

HOW TO REACH THEM

Address

Flambeau, Inc.,
Attn: Mail Code NPT,
P.O. Box 97,
Middlefield, OH 44062

Phone Number

(440) 632-1631 (For submission questions)
Fax: (440) 632-1581

Email

info@flambeau.com (Not for submissions)

Some of their products
Flambeau fluid systems: Plastic tanks, Washer bottles, Caps & reservoirs for automobiles and other systems, and more.

Ornamates™: Hunting decoys, Mailboxes, Mailbox accessories, Ornaments for lawns, ponds and gardens and more.

Flambeau Outdoors: Reels, Jig heads, Worm weights, Soft tackle systems, Fly fishing, Kids fishing gear, Tackleboxes, Decoys, Accessories, Calls, Gun cases, Bow cases and more.

Art Bin®: Art Totes and storage boxes, Craft storage boxes, Scrapbooking storage totes, pencil boxes and more.

Duncan® Toys: Yo-Yo's, Juggling products, Foot bags, Yo-Yo trick book, and more.

Flambeau Hardware: Hardware storage and totes, and more.

Flambeau Medical: Paramedic cases, Gear boxes, First Aid cases, Instrument boxes and more.

Flambeau Industrial & Packaging: Conductive storage bins, Conductive cases,

Storage packs, Utility boxes, Tailgators™, Step n Stor and more.

Some of their brands
ArtBin® Art & Craft Products, Duncan® Toys, Flambeau Hardware™ Products, Flambeau Outdoors™ Products, Flambeau Premiums & Special Markets, OrnaMates™, Vlchek® Floral Containers

Idea submission and submission policy
Flambeau says they accept and encourage the submission of outside ideas, inventions, and/or information regularly. If you have a product idea or invention you would like to introduce to them, fill out their online Inventor's Submission Release & Questionnaire.

To locate the questionnaire on their website, go to Flambeau's *Home page.* Under the *Corporate Information* link, go to, *Contact Us*. Scroll down to Inventor's Submission Form heading.

When filling out their online form, be sure to read all details and instructions included on the form. When the form is complete you will be able to submit the information via e-mail using the E-mail Form button. Once that process is finished, you can include additional comments and attach all pictures, drawings or other information pertinent to your invention.

If submitting your idea via mail or fax, fill out the form and print. Sign the release and send any other pertinent information along with your form to the above address or fax number.

Comments

Go to Flambeau's *Product Division* heading for links to their various websites showcasing the products listed above.

Question:

How would I pursue a patent or copyright focused on creating a game that ties in with a well known book series and movie?

Answer:

If you are going to use the name of the book/movie for your game, you would need to contact the individual or company that owns the rights to it and request permission. Use is usually granted in the form of licensing which would give you the right to use their name, their design, etc. for your game.

If your game does not require the use of the name, created characters from the series or movie, or anything that could be construed as an infringement on the creator's rights, you will not need permission and can register a copyright or file for a patent through the USPTO, or hire a professional to protect your game idea.

Fundex Games

Stuff they sell
Indoor and Outdoor games for children and adults

Website
Home page: http://www.fundexgames.com

Inventor page:
http://www.fundexgames.com/inventors.php

HOW TO REACH THEM

Address
Fundex Games
Attention: Product Development Manager
1570 S. Perry Road
Plainfield, IN 46168

Phone Number
(800) 486-9787
Fax: (317) 248-1086

Email
Website email form available

Some of their products
Alex Beard Face to Face, Bowling Dice, Phase 10, Don't Tip the Waiter, Booby Trap, Bunco Party in a box, Sketchy, Privacy, Heist, Hit the

Deck, Double Dutch Pro, Chuck O Splash (pool game), Dive Pets, Toss Across Splash and more.

Idea submission and submission policy
Send a completed New Product Submission Form along with a sample of your game and game instructions to the above address.

Submission form can be found by going to the Fundex *Home page*, then to *Contact Us* link. Click on *Inventors* link to access the submission page.

Fundex says they periodically review New Product Submissions and will contact you if they are interested in developing your product idea.

Garden Weasel

Stuff they sell
Garden Weasel is currently seekir
Related Items, Gift Items, Lawn ar
Outdoor living, Hardware, House W
Improvement, Automotive, Garment Care,
Cleaning.

Website
Home page: http://gardenweasel.com/

Submission form page:
http://gardenweasel.com/inventors.html

HOW TO REACH THEM

Address
Garden Weasel
Dept.WWW
1025 W. 8th Street
Kansas City, MO 64101-1200

Phone Number
N/A

Email
info@gardenweasel.com

Some of their products
The Garden-Weasel®, Garden Claw®,

eedPopper®, RuXXac®Cart, and TRAPP®Candles

Idea submission and submission policy
Go to home page and click on *inventors click here,* for submission information, or go to submission link above. This will take you to the required online submission forms.

Comments
All of Garden-Weasel's® products are affiliated with their company through acquisition, exclusive license, strategic alliance, or inventor license.

Do not submit information on an idea. Only submit information about a product that is secured by patent protection or is patent pending.

GCI Outdoor

What they sell
Portable Recreation Gear

Website
Home Page: http://www.gcioutdoor.com/

Submission Page:
http://www.gcioutdoor.com/inventors_corner.html

HOW TO REACH THEM

Address
GCI Outdoor, Inc.
66 Killingworth Road
Higganum, CT 06441

Phone Number
Main Office: (860) 345-9595
Toll Free: (800) 956-7328
Fax: (860) 345-2966

E-Mail
info@gcioutdoor.com

Some of their products
Sitbacker™ Canoe Seat, Packseat™ Portable Stool, Pico Arm Chair™ Telescoping Directors Chair, Wilderness Recliner™ Luxurious Recliner, Little Titan™ Kids Chair,

Bleacherback™ Stadium Seat and more.

How to submit your idea At *Home Page*, go to *Inventor's Corner* link at the bottom of the page or go directly to submission page link.

Fill out the website form and a representative from the New Product Development Department will send you a new product submission letter.

Provide all materials, including all patents or patent application numbers that support your new product submission.

GE

Stuff they sell
Appliances, Electronics, Aviation, Electrical Distribution, Healthcare, Lighting, Oil and Gas, Rail, Security, Water

Website
Home page: http://www.ge.com/

Idea submission info page:
http://www.ge.com/contact/submitted_ideas/index.html

HOW TO REACH THEM

Address
GE Corporate Headquarters address:
General Electric Company
Fairfield, CT 06828

Phone Number
(203) 373-2211

Some of their products
Computer Accessories: Mouse pads, Web cams, Headphones, Headsets, Keyboards, Speakers, Mice.

Home Electric Products: Lighting fixtures, Timers, Door chimes, Book lights.

Electronic accessories: Digital converter boxes, Antennas, Cordless phone batteries, Blue tooth.

Home appliances: Ranges, Refrigerators, Washers, Dryers.

Other: Turbojet engines, Gas turbines, Metering systems, Generators, Water purification technologies, Security systems, Fluorescent bulbs, Mammography, Digital radiography, wind turbines, Entellisys™ Low Voltage Switchgears, and more.

Idea submission and submission policy

Go to submission page listed above or (1) Go to *home page* (2) Click on *contact information* at bottom of page (3) Click on *submitting ideas and inventions.*

Read instructions and follow the appropriate links to make a submission. Submissions are made via internet webpage.

Comments

If the submitted ideas are original and new to GE and are adopted for use, an honorarium will be awarded to the submitter. The amount of the honorarium will be established according to the company's judgment, but will not exceed $5,000.

General Mills

Stuff they sell
Food products

What they are looking for
General Mills is interested in technologies that would apply to one of their current lines of business: Baking Products (Brownies, Cakes, Frosting, etc.), Frozen Vegetables, Frozen Pizza and Snacks, Frozen Pastries, Cereal, Refrigerated and Frozen Dough (Biscuit, Cookie, Bread, Rolls, etc.), Shelf Stable Meals, Meal Kits, Soups and Side Dishes, Yogurt and Yogurt Beverages, Snack Bars, Fruit Snacks, Popcorn, Salty Snacks, Soy Beverages.

Website
Home page: http://www.generalmills.com/

Overview for G-WIN submission:
https://openinnovation. generalmills.com/

HOW TO REACH THEM

Address
N/A

Phone Number
(763) 764-4946

Email
g.win@genmills.com

Some of their products
Apple Cinnamon Cheerios, Bac-os Bits, Bagged fruits, Bagged vegetables, Bowl Appetit cheddar broccoli pasta, Biscuits buttermilk, Boo Berry, Breadsticks, Brownie mixes, Hamburger Helper lasagna, Honey nut clusters, Sweet rolls orange, with Icing, Milk and cereal bar Cocoa Puffs. For a complete list, view product index on *home page*.

Some of their brands
Betty Crocker, Bisquick, Gold Medal, Pillsbury, Cascadian Farm, Latina, Totinos, Progresso, Fruit roll ups, Fruit Gushers, Wheaties, Cheerios, Chex, Pop Secret, Nature Valley, Yoplait, Haagen Dazs, Green Giant, Old El Paso, Gardettos.

Idea submission and submission policy
Go to link for Overview: https://openinnovation.generalmills.com/

This G-Win website will list current innovations sought after by General Mills with links to submit your proposal for these products.

G-Win's online system will walk you through the submission process. You will be sent a confirmation that your submission has been received and will be contacted within 2-3 weeks after the preliminary review of your submission.

Comments
General Mills can accept a variety of non-confidential forms of information but requests that the proposal be based on a product, package or process innovation technology.

While patented technologies are preferred, a patent is not required to submit your idea to General Mills.

Accepted idea submissions will be implemented via one of the following: licensing agreement, manufacturing, investment, joint venture, proof of concept, supply agreement or other.

Graham Beauty

Stuff they sell
Innovative beauty products for salon, spa, nail and barber professionals

Website
Home Page: http://www.grahambeauty.com/

HOW TO REACH THEM

Address
Graham Beauty
A Division of Little Rapids Corp.
2273 Larsen Road
Green Bay, WI 54303 USA

Phone Number
(920) 490-5349 (Gerry Paul)

Email
grahambeauty@littlerapids.com

Some of their products
U-be Hairweave Cap, Sanek®See thru Foil, Neck Essentials™ Style Strips, Spa Essentials® Eye/Lip Applicator, WUBBIES® Embossed Towels, BARBEE®Towels, Spa Essentials® Paraffin Strips, Birchwood Applicators and much more. See website for full line of products.

Idea submission and submission policy
Contact their Product Development Manager, Gerry Paul at (920) 490-5349 or at: gpaul@littlerapids.com

Guthy-Renker

What they do
Guthy-Renker is direct response television marketing.

Stuff they market
Automotive, Beauty, Fitness, Health & Nutrition, Entertainment, House wares, Electronics, Golf & Outdoors, Personal Development & Business Opportunities, Other

Website
Home page: http://www.guthy-renker.com/

HOW TO REACH THEM

Address
Att: Product Submissions
Guthy-Renker Corporation
41-550 Eclectic, Suite 200
Palm Desert, CA 92260

Phone Number
(760) 773-9022
Fax: (760) 773-9016

Email
productsubmissions@guthy-renker.com

Some of their products
Proactiv Solution, Principal Secret, Winsor Pilates, Meaningful Beauty, Get the Edge, Natural Advantage, Youthful Essence, Core Secrets, Sheer Cover, In An Instant, Zumba Fitness, The Dean Martin Celebrity Roasts.

Ideas submission and submission policy
(1) Go to *home page* (2) Click on, *About GR* (3) Go to *submit a product*.

Read criteria and instructions. Choose appropriate form, print and fill out. Send completed form with requested information/samples to the above address.

Comments
Guthy-Renker is a direct response television company (DRTV).

To qualify for Guthy-Renker, your product must

(1) Not yet be sold at retail

(2) Be extremely visual and demonstrable

(3) Be beneficial to consumers and offer genuine value.

(4) Be impulse driven.

(5) Have the ability to entice repeat buys and backend/continuity sales.

(6) Have mass market appeal in multiple demographics.

Henkel

Stuff they sell
Laundry & Home Care Products, Cosmetics & Toiletries, Adhesive Technologies

Website
Home Page: www.henkelna.com

Innovation page:
http://www.henkelna.com/innovation/henkel-innovation-partnership-program-6085.htm

HOW TO REACH THEM

Email
Use the *Contact Us* link and click on *Corporate R&D and Innovation Topics* link. Submit your email questions or comments on this website email page.

Some of their products
Renuzit®Air Fresheners, Roller Scents™, Purex®, Soft Scrub®, Combat® insecticides and much more.

What they are looking for
Henkel is interested in ideas for products, processes and designs in the following three categories:

Laundry & Home Care: Heavy duty detergents, bath and toilet cleaners, glass cleaners, air fresheners, floor and carpet care products, kitchen cleaners, special purpose cleaners, dishwashing detergents, scouring agents, and insecticides for household use.

Cosmetics & Toiletries: Hair washing and hair care products, bar and liquid soaps, shower gels, body washes, bath additives, skin creams, deodorants, skin care products and products for professional hair dressers.

***Technologies**:* Wallpaper pastes, renovating products, tiling adhesives, sealants, assembly adhesives, contact adhesives, glue sticks, roofing products, correction products, adhesive tapes, wood adhesives, cleaning agents and more.

Idea submission and submission policy

Go to *Home Page* and click on the *Innovation* link, then go to *Henkel Innovation Partnership Program* in the left column.

The sub links are informative and important to read before moving onto the submission page.

Once you are ready to proceed, go to *Submit an Idea*. This will lead you through the initial process for submission.

Comments

You need to have a patent, a published patent application, or a registered and published utility model or design before you submit your product idea. Review of your product may take several months.

Hog Wild Toys

Stuff they sell
Toys, Games, and Unique Gift Items

Website
Home page: http://www.hogwildtoys.com/

Inventor's page:
http://www.hogwildtoys.com/inventors.html

HOW TO REACH THEM

Address
Hog Wild Toys
221 SE Main Street
Portland, Oregon 97214

Phone Number
Toll free (888) 231-6465
Fax (503) 233-0960

Email
invent@hogwildtoys.com

Some of their products
The Benders, Temperature Controlled Shower Light, Faucet Light, Peeramid Book Rest, Robot Calculator, Robo Vacuum, Zoo Pops, Dino Puzzle Cookie Cutter, Spyrogyro, Locket Pen, Tape Gun, Fridge Pins, Ice Tea Mixer,

Dino Sticks, Snap Watch Series II, Rubber Band Shooter, Power Popper, Acrobots, Candy Popper, Scrap Metal Sculpture. See website to view all of their products.

Idea submissions and submission policy
Go to inventors page information link above or (1) Go to *home page* (2) Go to *inventors* link and read what they are about.

Contact by email for Non-disclosure agreement (NDA), sign and return and if Hog Wild is interested, they will want to see a prototype.

Comments
Hog Wild is looking for freethinking toys and gifts. They pay competitive royalties and will sign and honor all Non-disclosure agreements. If you have questions, just email.

Homax Products

Stuff they sell
Home Improvement Products for DIY: Paint Accessories, Kitchen/Bath, Fencing, Surface Prep, Odor/Moisture, Texture, Tile Flooring, Tarp/Tie Down, Patch Repair, Cleaners/Removers

Website
Home page: www.homaxproducts.com

HOW TO REACH THEM

Address
The Homax Group
3701 Shoreline Drive #202C
Wayzata , MN 55391

Phone Number
(952) 471-9009 Phone
(952) 471-9004 Fax

Email
homax@homaxproducts.com (Customer service)

Some of their products
Waste Away paint hardener, Lead test kit, Brush and roller cleaner, Paint guides, Corner paint guides, Caulk finisher, Tub and tile brush

on finish, Caulk remover, Appliance touch up paint, Splash guards, Fixture trim, Shower Squeegees, Odor absorbing gels, Odor eliminator, Fan filters, Odor air magnet, Drywall taping tool, Wall guard, Crack patch, Nail hole patch, Popcorn texture, Knock down texture, Paint 'n Tex, Hopper gun system, Fiberglass cleaner, tile grout coating. For complete list, see website.

Some of their brands
Homax, Tile Guard, Bix, Magic, Jasco, Rhodes American, Gonzo, Goo Gone, Natural Magic, Myro, and OOPS!

Idea submission and submission policy
Homax requires that only products that are patented or patent pending and preferably already in the public domain be submitted. If you are already showing your product publicly they would be happy to review it. If you think you need more protection to show it publicly, please wait until your product is securely protected before submitting.

Contact Homax to request an Idea Submission Policy. Once that is completed, sign and return it along with the description of your product to the above, Minnesota address for the first step in the review process.

Trade Shows & Your Product

Exhibiting your product at trade shows is a great way to get your idea picked up by manufacturers or attract the attention of well known direct marketers like HSN and QVC.

The catalyst for Debbie Meyer's success, owner of Housewares America, was when her product invention was discovered by QVC at an international home and house-wares show. This debut invention, known as the Kake-Kut'r, is designed to neatly slice pieces of cake and serve them up.

Currently Debbie Meyer launches all of her products on HSN. House-Wares America, Inc. brings in revenues of $100 million annually.

Home Shopping Network (HSN)

What they do
HSN describes itself as an interactive lifestyle network and retail destination.

Stuff they sell
Jewelry, Fashion, Handbags, Shoes, Beauty, Kitchen and Dining, Electronics, Home Décor, Home Solutions, Wellness.

Website
Home page: http://www.hsn.com/

HOW TO REACH THEM

Address
HSN
PO Box 9090
Clearwater, FL 33758

Phone Number
N/A

Some of their products
Scarves, Boots, Belts, Accessories, Shirts, Jeans, Pots and Pans, Pressure cookers, Cutlery, Cookbooks, Computers, Video games, Printers, Scanners, Televisions, Scrapbooks

supplies, Make-up, Hair care, Perfumes, Bracelets, Charms, Pendants, Watches, Ellipticals, Strength training, Treadmills, Artwork, Bed & Bath, Automotive care, and much more. See website.

Idea submission and submission policy
(1) Go to *home page* (2) Click on *become an HSN partner* (3) Follow *Application Process* link and *Terms and Conditions*. These will instruct you on HSN's policies before you submit.

Comments
The following should be included in the application: Product description and your target customer, photo or brochure of product that can be uploaded, quantity available, and the suggested selling price.

HSN is there to distribute your product, however if you do not have your item manufactured, they may be able to direct you to a manufacturer.

Hussmann

Stuff they sell
Refrigeration Systems, Walk in Coolers, Food Service Display Cases and their Components

Website
Home page: http://hussmann.com/

Submission page info:
http://techportal.hussmann.com/idea%20generation/Pages/default.aspx

HOW TO REACH THEM

Address
Hussmann World Headquarters:
12999 St. Charles Rock Road
Bridgeton, MO,
63044-2483
USA

U.S. Executive Offices:
Ingersoll-Rand Company
Corporate Center
155 Chestnut Ridge Road
Montvale, NJ 07645

Phone Number
(314) 291-2000 (General inquiries)

Some of their products
Always Bright, LED lights for shelves of multi-deck medium temp display cases and for narrow footprint (RLN) Reach-Ins with Innovator Doors, Low temp Excel islands, Integrated night curtains, Always Clear no fog door system, Hot food countertop merchandiser, Self contained deli food merchandiser and much more. See site for full list.

Idea submission and submission policy
Go to the submission page link listed above, or go to the Hussmann *home page* (also above) and click on *submit a new product idea* link.

Submit your idea to Ingersoll Rand Climate Control Technologies via web submission page.

Comments
Your product idea can be new to Hussmann, new to customer market, or be an improvement or a replacement to an existing product.

Jada

Stuff they sell
Toys, Radio Control Vehicles, Models

Website
Home Page: http://www.jadaclub.com/

Submission Page:
http://www.jadaclub.com/inventors.php

HOW TO REACH THEM

Address
Jada
938 Hatcher Ave
City of Industry, California 91748

Far East Branch Office:
3F, Tower B,
New Mandarin Plaza,
Kowloon, Hong Kong.

Some of their brands
Snap 'N Build™, Dub® City®, Speed Racer ™, Snap Shots™, and more

Idea submission and submission policy
(1) At *Home Page*, go to *About Us* link and (2) Click on *Inventors.* Or Go directly to submission page listed above. Fill out Jada's

online submission form at their submission page and send in your idea.

Jarden Home Brands

Stuff they sell
Home products such as Canning jars and products, Fire Logs, Toothpicks, Playing cards, Other

Website
Home page:
http://www.jardenhomebrands.com

HOW TO REACH THEM

Address
Jarden Home Brands
14611 W. Commerce Road
Daleville, IN 47334

Phone Number
(800) 392-2575

Email
info@jardenhomebrands.com

Submission email:
JHBDiamondInfo@jardenhomebrands.com

Some of their products
Matches, Lighters, Toothpicks, Clothespins, Plastic cutlery, Straws, Fire logs, Canning jars and lids, Pectin, Fruit fresh, Plastic freezer jars,

Mason jars, Home canning accessories, Playing Cards, and more.

Some of their brands
Diamond®, Pine Mountain®, Java Log®, Ball®, Bernardin®, Kerr®, Northland®, Bicycle

Idea submission and submission policy
Contact Jarden using the submission email and your product ideas will be forwarded to their marketing and research & development teams.

Joan Lefkowitz's *Accessory Brainstorms*

What they Do
Licensing Agent, Marketer, and Consultant for Fashion and Beauty Accessories

Website
Home Page:
http://www.accessorybrainstorms.com/

HOW TO REACH THEM

Address
Accessories Brainstorm, Inc.
389 Fifth Avenue Ste 705
New York, NY 10016

Phone Number
(212) 379-6363

Email
Use their *contact us* page to email

Categories they are looking for
The invention categories *Accessories Brainstorm covers* include fashion jewelry, hair inventions, headwear and neckwear, weather related accessories, handbag items, body-wear, lingerie accessories, personal care

inventions, footwear, beauty inventions as well as lifestyle inventions and products that make life easier.

Products are developed for fashion, cosmetic, specialty chain, drug, spa, salon, mail order catalogues, television retailers, and infomercial companies.

Some of their products
Whirl a Bun, Lap Top Manicure Tray, Tag Tamers ™, REM Spring, and more

Idea submission and submission policy
Contact Accessories Brainstorm at the above address, phone number or email contact page for further information.

Comments
Accessory Brainstorm receives compensation on a contingency basis for sales and licensing. They don't get paid unless you do.

Consultations are fee based.

Johnson & Johnson

What they do
Johnson & Johnson's COSAT seeks scientific breakthroughs. They explore technologies such as molecular diagnostics, cell therapies, nutritionals, medical devices, health care informatics platforms, and bio-artificial organs.

Website
Submission/home page:
https://www.jnjcosat.com/

Johnson & Johnson home page:
http://www.jnj.com/ connect/

HOW TO REACH THEM

Address
Johnson & Johnson Services, Inc.
Corporate Office of Science & Technology Division
410 George Street
New Brunswick, NJ 08901-2021

Phone Number
N/A

Email
COSAT@CORUS.JNJ.COM

Some of their products/brands
The Clean & Clear® Advantage® Acne Control Kit, Cortaid® Poison Ivy Toxin Removal Cloths, Desitin® Clear™ Multi-Purpose Ointment, Johnson's® 2-in-1 Shower and Shave 24 Hour Moisturizing Wash, Johnson's® Baby Oil Body Wash, Johnson's® Cucumber Melon Baby Wash, Baby Lotion and Baby Powder, Johnson's® Shea & Cocoa Butter Baby Cream, Listerine® Whitening® Quick Dissolving Strips, Splenda® No Calorie Sweetener Minis, Positively® Ageless™, and much more. See website for list.

Idea submission and submission policy
(1) Go to Submission/*home page* listed above and (2) click on *submit general ideas to COSAT* link. (3) Click on *Submitting your idea* and create a user profile before you submit your idea.

Johnson & Johnson will review only United States Issued Patents or official patent publications. In addition, these patented ideas must fall within areas of strategic interests to Johnson & Johnson.

They will not accept for review unpatented ideas, advertising slogans or promotional programs, unpublished patent applications, prototypes or models.

Comments
Although Johnson & Johnson seeks a broad area of science and technology, the following is where there is a large unmet need:

Technology Areas:
Biotechnology
Cell Therapy
Diagnostics / Biomarkers
Drug Discovery Technology
Electro-stimulation / Neuro-stimulation
EMR/ PHR/ Healthcare Delivery
Energy-Based Therapeutics
Gene-Targeted Therapies
Home Monitoring
Intelligent Devices
Minimally Invasive & Computer Assisted Surgery
Microsystems (MEMS / Nanotechnology)
Nutraceuticals and Natural Products
Regenerative Medicine

Therapeutic Areas:
Cardiovascular
Central Nervous System
Gastroenterology
Hearing
Immune / Inflammatory
Infection Control
Metabolic
Oncology
Ophthalmology

Pain Management
Pulmonary
Medicine
Renal
Urology & Sexual Health

Jokari

Stuff they Sell
Solution Gadgets for Kitchen and Home, Storage and Organization Products

Website
Home Page: http://www.jokari.com/

HOW TO REACH THEM

Address
Jokari
1815 Monetary Lane #100
Carrollton, TX 75006

Phone Number
(972) 416-5202
Toll-free (800) 669-1718

Some of their products
Back Pack Rack, Purse Rack, Ultimate Funnel, Burger Maker, Stor Pods, Shelf Pods

What they are looking for
Jokari is currently looking for innovative ideas in the house wares category. If you have an idea or a product that is a kitchen gadget, storage and organization product or other household solution gadget, you may submit it to Jokari for review.

Idea submission and submission policy
Go to *Home page* then to *inventors* link and submit your idea via the provided online form.

Josten

Stuff they sell
Graduation Products

Website
Home page: http://www.jostens.com

HOW TO REACH THEM

Address
Corporate Headquarters
Jostens, Inc.
3601 Minnesota Drive
Minneapolis, MN 55435

Phone Number
(952) 830-3300

Some of their products
Rings, Jewelry, Yearbooks, Caps, Gowns, Diplomas, Super Bowl rings, Fantasy sports, rings, Sterling silver cheerleading jewelry collection, Trophies, Senior apparel, including Hoodies, Shorts and Pants, Keepsakes, Memory books, Albums and more.

Idea submission and submission policy
(1) Go to *home page (*2) Click on *about us* (3) Click *idea submission* link and read Josten's

submission policy and agreement prior to submitting idea.

Jostens requires that your submission be in writing so that the company can properly communicate and consider the idea.

Submissions will be accepted via their new online idea submission process.

Kapro Industries, Ltd.

Stuff they sell
Manufacturer and developer of innovative hand tools for the professional market: Levels, Lasers, Layout, Marking and Measuring, Merchandising and Promotions

Website
Home page: http://www.kapro.com/

Submission page:
http://www.kapro.com/ino_idea.asp

HOW TO REACH THEM

Address
Kapro Industries Ltd.
Innovation Manager
Kadarim 12390
Israel

Phone Number
N/A for submissions

Email
innovation@kapro.com

Some of their products
Gradient levels, Construction levels, Tool box levels, Post and pipe levels, I-Beam levels,

Laser accessories, Rulers and Straightedges, Framing squares and Drywall, Squares and Angles, Chalk Lines, Measuring tapes and Wheels, Countertop displays, Freestanding floor displays and more.

Some of their brands
Zeus ™ 990, 985D Digiman ®Digital Level, 893 T-Laser™, Prolaser®Laser, Plumb Site®, Dual View™

Idea submission and submission policy
Go to submission page link above, or (1) Go to *home page* (2) Go to *innovation center* link and click on *idea submission*. Download, read and sign Submission Agreement and Submission Form and send to Innovations Manager. If you have any questions, submit them through the email link above.

Kapro does not accept email submissions. They will do their best to respond within 60 days of receiving your completed forms.

Comments
Kapro is always developing and introducing new innovative hand tools, and welcomes suggestions, ideas, and recommendations for new products, new features, and product improvements. For General information about Kapro, go to quality@kapro.com

Kellogg

Stuff they sell
Food Items

Website
Home Page:
http://www2.kelloggs.com/greatideas/

HOW TO REACH THEM

Address your questions and concerns at Kellogg's email page by using the *Contact Us* Link or by going to:

http://www2.kelloggs.com/ContactUs.aspx
If your questions cannot be addressed via email, call or write...

Address
Kellogg Consumer Affairs,
P.O. Box CAMB
Battle Creek, MI 49016

Phone Number
(800) 962-1413

Some of their products
Pop Tarts®, Cheez-Its®, Rice Krispies, Nutri-Grain®, Keebler® EL Fudge, Famous Amos®, Eggo®Waffles, Special K Crackers, Eggo®

Syrup, Ready Crust®, Corn Flake Crumbs, Morningstar Farms®Natural and Organic Veggie Foods, and much more.

Idea submission and submission policy
If you have a new product that is partially or fully developed and could be ready to launch quickly or you have a proposed business collaboration or a patented food, packaging or processing technology, go to *home page* and click on the *Great Innovation* link. Follow the prompts for submission.

Kellogg will then review your submission and consider their interest. You should be contacted within 6-8 weeks.

Comments
Very Important: If your submission falls under the Big Idea category, Kellogg will not compensate you. Most of us don't want to give our ideas away for free, so consider having your idea "partially or fully developed" for a quick launch. This way it can be submitted under the Innovation link and you can profit from your labor.

Although you don't need a patent or copyright protection on your innovation, discussing it with Kellogg may affect your rights. It would be wise to seek an attorney prior to disclosure.

Review their site for any further information you may be looking for.

Kimberly Clark

Stuff they sell
Health and Hygiene/Paper Products
(Note: Currently K-C is working with brokers/agents to connect with independent inventors)

Website
Home page: http://www.kimberly-clark.com/

HOW TO REACH THEM

Address
Kimberly-Clark Corporation Dept: GATA
P.O. Box 2020
Neenah, WI 54957-2020 U.S.A.

Phone Number
(800) 789-4487

Email
alliances@kcc.com

Some of their products
Diapers, Tissues, Feminine hygiene products, Toilet paper, Paper towels, Napkins, Flushable wipes, Kleenex® Moist Cloth, Kleenex® Splash 'n Go®, and more.

Some of their brands
Kleenex®, Kotex®, Viva®, Scott®, Depend®,

Fiesta®, Kimcare®, Pull Ups®, Little Swimmers®, KleenGuard®, Kimberly Clark Professional®, Huggies®, Poise®, Hakle®, Cottonelle®, Good Nites®, Scottex®, Snugglers®, and more.

Idea submission and submission policy
For more information on partnering with Kimberly Clark, (1) go to home page, (2) go to *about us* and click on *innovations* (3) go to *innovating with K-C* link.

K-C is presently working with brokers/agents to connect with independent inventors.

No Sold Thru the Band

Kraco Enterprises, Inc.

Stuff they sell
Automotive Aftermarket, Related Home, Workshop or Hardware Products.

Website
Home Page: http://www.kraco.com/home.htm

Submission page:
http://www.kraco.com/ideas.htm

HOW TO REACH THEM

Address
Kraco Enterprises, Inc.
505 E. Euclid Avenue
Compton, CA 90224

Phone Number
Phone: (310) 639-0666
Fax: (310) 604-9838
Toll Free (800) 678-1910

Email
N/A for idea submission

Some of their products
Rubber floor mats, Specialty floor mats, Rubber runners, Rubber cargo liners, Fade

resistant utility mat, Vehicle carpet mats and more.

Idea submission and submission policy
Go to submission link above or (1) Go to *home page* (2) Click on *ideas wanted.*

After reading their instructions, if you are interested in submitting new product concepts or prototypes for review and evaluation, please send information to the above address.

Comments
Products are wanted for development, license and retail distribution. You must have at least a provisional patent registration for your product idea to be evaluated.

Kraco is looking for:

Ecologically responsible products:
Products that are recycled, use a high content of post consumer waste, products that use a high content of post production waste. Their main focus is on auto, garage, and home flooring products.

Protection:
Protective items for auto surfaces and garage. This could include innovative seat covers, cargo, steering wheel covers, sun shades, organizers for auto, garage organization or

home floor protection such as innovative mats, tiles, flooring, etc.

Pride of ownership: Which would be products that relate to auto and home flooring, but are special enough for the people who care significantly about how the products look as well as how well they protect their auto, home flooring or garage.

Question:

I really want to try to sell this special sauce that I make, but I have no idea how to do that. What steps do I need to take? Since it is a food product I think I would have to go through the FDA, but how would I do that?

Answer:

If you are looking at selling it in small amounts where you are making and bottling it yourself or you want to open your own manufacturing plant for the product, you should contact the health department in your state and they will be able to give you the low-down on what the legal requirements are. If you plan on someone else manufacturing it for you, it is as simple as contacting food manufacturing companies and setting up an agreement with them to manufacture your product. They would be responsible for following all state and FDA requirements.

You can shop around for costs and quality of various manufacturers before you agree to anything and use that data to determine your wholesale costs and expenses before presenting your product to various outlets (stores, catalogues, etc.). Next, you would want to create a sales pitch, sample product and information. These you would take with you on the appointments you schedule with buyers/purchasing agents to market your product. You might also consider pitching your idea to restaurants that may want to enter into an exclusive contract with you to use and/or sell your sauce.

Last, you can also sell your idea outright or license it and earn from royalties. There are many well known food manufacturers that are interested in purchasing new products from outside sources.

Kraft Foods

Stuff they sell
Food products

Website
Innovation home page:
www.innovatewithkraft.com

HOW TO REACH THEM

Address
Kraft Foods
Steven Goers VP
Open Innovation
1 Kraft Court
Glenview, IL 60025

Phone Number
(866) 676-4332

Some of their products
Crackers, Cheese, Juice, Coffee, Nuts, frozen pizza, Cookies, Salad dressing, Macaroni and Cheese, Bacon, Cream cheese, Nuts…

View http://www.kraft.com for an extensive list of their products and brands.

Some of their brands
A-1, Alpen Gold, Capri Sun, Cheez Whiz, Chips Ahoy, Cool Whip, Cracker Barrel,

California Pizza Kitchen, Crystal Light, Di Giorno, Gevalia, Maxwell House, Honey Maid, Handi Snacks, Ritz, Royal, Velveeta, Philadelphia, Planters, Wheat Thins, Tang, Miracle Whip, and much more.

Stuff they are looking for:

Technology: Weight Management, Natural Preservation, New Flavor Systems.

Packaging: Improve sustainability, Insure product safety, Increase consumer convenience.

Products: Health and wellness/Healthy foods, Quick meals, Snacks, Premium (restaurant quality food).

Processes: Improving microwave cook time, General innovations that simplify food preparation, Non-destructive detection of non-ferrous foreign materials, High speed detection of leaks in packaging, Reduction of energy use in food manufacturing, Reduction in water use in food manufacturing, Reduction of waste materials in food manufacturing, and Recycling / reuse of manufacturing process waste.

Idea submission and submission policy
Go to the listed innovation page above and after you choose the region you live in, click on *submit your innovation* to review the procedures and acquire an idea submission form. You can also download a printable version or request one using the listed phone number.

Depending on the type of idea you have, you may have proprietary rights in it. Therefore, you should consider protecting it through patents or copyrights before submitting it to Kraft.

Kraft requests the following:

1. Please read and acknowledge your acceptance of terms of the Submission Agreement by signing a hard copy of the Submission Agreement.

2. Complete the brief "Idea Submission Form", which provides information necessary for Kraft to do an initial evaluation of your idea.

3. Mail the signed Submission Agreement, Idea Submission Form and any additional documentation you feel is necessary to provide to Kraft, to:

Steven Goers, VP Open Innovation, Kraft Foods, One Kraft Court, Glenview, IL. 60025

4. Once Kraft receives the signed Submission Agreement and Idea Submission Form, your idea will go to the appropriate Kraft person to assess their level of interest. If they need additional information, and you consider the additional information to be confidential, Kraft may enter into a written confidentiality agreement with you at that point to obtain it.

(*They note that most of the time, non-confidential disclosures are sufficient for Kraft to determine its level of interest, so in the vast majority of situations, confidentiality agreements will not be necessary).*

5. Kraft will try to report back to you with the results of its evaluation in about six weeks, although some evaluations take longer. Under no circumstances will Kraft be obligated to reveal the details of its evaluation.

6. Personal interviews are not necessary in most cases, but if an interview becomes necessary, it will most likely take place at a Kraft corporate location in the New York, NY, Chicago, IL or Madison, WI areas.

Comments

Kraft is looking for products that are ready to

be brought to market or can be brought to market quickly. Do not submit recipes, entertainment ideas, ideas for line extensions for existing Kraft products or packages, ideas for advertising/ promotions, etc. Such ideas fall outside the scope of this policy and will not be reviewed by their Innovations Team or considered for compensation.

If Kraft is interested in using your idea and the idea is protected or legally able to be protected by a patent or copyright, they may negotiate for license rights. Your compensation will be determined as a part of those negotiations.

If Kraft is interested in using your idea and it is not protected, or cannot be legally protected by a patent or copyright, but *is* new to Kraft and they adopt it, Kraft *may* grant you a nominal award, not to exceed $5000.

Kraftmaid Cabinetry

Stuff they sell
Glass Inserts, Moldings and Accents, Hardware, Cabinets, and Cabinet Innovations, Food Storage, Table Ware Storage, Prep-Storage, Cooking Storage, Clean Up Storage, Home Office and Kitchen Message Center, Bath Storage, Other.

Website
Home page: http://www.kraftmaid.com/

HOW TO REACH THEM

Address
Edgar A. Zarins
Intellectual Property Counsel
Masco Corporation
21001 Van Born Road
Taylor, Michigan 48180

(Submission address)

Phone Number
(888) 562-7744 (General questions)

Email
CustomerCarePPG@kraftmaid.com

Some of their products
Knobs, Pulls, Companion groups, Kitchen cabinets, Bathroom cabinets, and more.

Some of their brands
Harmony™ Storage

Idea submission and submission policy
Submit your request for consideration in writing to the above address.

Lancaster Colony

Stuff they sell

Glassware, Candles, Specialty Foods, Commercial Products, Automotive, Other (Contact Lancaster Colony for specific needs)

Website

Home page: http://www.lancastercolony.com/

HOW TO REACH THEM

Address

Att: Pat Schnieder
Lancaster Colony Corporation
37 W. Broad St.
Columbus, OH 43215

Phone Number

(614) 224-7141

Email

Emails can be sent by going to *Company Information* link, then to *Contact us.*

Some of their products

Texas Toast croutons, Jack Daniels®mustard, Egg noodles, Dressings, Dips, Desserts and Glazes, Caviar, Frozen foods, Frozen garlic bread, Yeast rolls, Candles.

For Hotels: Plastic Ware, Coffee Urns, Matting, and more.

Some of their brands
New York®, Marzetti®, Romanoff, Mama Bella®, Inn Maid®, Candle-Lite®, Wescon®

Idea submission and submission policy
At *home page*, go to *FAQ* link. Click on the word *policy*, located in the body of their answer to access submission policy page.

Review and execute their policy agreement and send it to the above address along with your idea submission.

Lancaster says: Ideas that are not covered by patents or copyright will be considered only with the understanding that the use for such ideas, including communication to other organizations for evaluation, are matters of discretion to be determined solely by Lancaster Colony.

Patented or copyrighted ideas will be considered only with the understanding that the submitter will rely on such rights as he may have under the patent and copyright laws for protection. Pending applications for a patent will be treated in the same manner as ideas not covered by a patent, unless and until a patent issues.

Lifetime Brands

Stuff they Sell
House Wares

Website
Home Page: http://www.lifetimebrands.com/

HOW TO REACH THEM

Address
Corporate Headquarters
Lifetime Brands
1000 Stewart Avenue
Garden City, NY 11530

Phone Number
(203) 594-8808

Email
wwtuttle@yahoo.com

Some of their Products
Bake Ware, Crystal, Cutlery, Home Décor, Pantry Ware, Mirrors, Kitchen Ware, Gift Ware, Flatware, Stemware, Entertainment, Frames, Dinner Ware, Cutting Boards, Bath Accessories, Sterling Flatware, Barware, and more.

Stuff they are looking for
House Wares and Tabletop Products

Idea submission and submission policy
To find out more about licensing your product idea, contact Warren Tuttle at the above email or phone number.

Lisle Corporation

Stuff they sell
Automotive Tools

Website
Home page: http://www.lislecorp.com

Inventor/submission information:
http://www.lislecorp.com/about/program/

HOW TO REACH THEM

Address
Lisle Corporation
P.O. Box 89
Clarinda, Iowa 51632-0089

Phone Number
Phone (712) 542-5101
Fax: (712) 542-6591

Email
info@lislecorp.com

Some of their products
Drill grinder, Magnetic plugs, Pipe thread, Deep reach assemblies, Metric thread, Window view gauges, Breather vents, Hex bit, Small master chuck key set, Flare nut wrench, Torque angle meter, Brake cylinder hone, Brake spring pliers, Tie rod separator, Shock

absorber tool, Ball joint separator, Oxygen sensor thread chaser, Valve core tool, Filter socket set, Battery handler, Battery brush, Computer safe circuit tester, and much more. A complete list of products can be viewed at their website under *browse tools.*

Idea submissions and submission policy
(1) Go to *home page* (2) Click on *idea program* link.

Fill out the online form and Lisle will send you an Idea Disclosure Agreement. After receiving disclosure agreement in the mail, return completed and signed agreement to Lisle along with a description of your tool idea.

Please note that all disclosures are made on a non-confidential basis.

Comments
Compensation will be in the form of a financial award, or a royalty agreement.

Maddak, Inc.

Stuff they sell
Independent and assisted living devices for kitchen, dressing, household, pediatrics, exercise, grooming, mobility, eye care, bedroom, educational, wheelchair accessories and more.

Website
http://service.maddak.com/

HOW TO REACH THEM

Address
Maddak Inc.
661 Route 23 South
Wayne, NJ 07470

Phone Number
(973) 628-7600 ext. 3262
Fax (973) 305-0841

Email
mmagnifico@maddak.com (submissions)
CustService@Maddak.com

Some of their products
Swiveling back scrubber, Swiveling lotion applicator, Denture safe cup, Baby booster infant stimulation kit, Maddacare children's

seats, Sit and ride ambulatory aid, Hand gym exercise unit, Bowling ball pusher, Building language through picture symbols, and much more. See online catalog at site for full product list.

Idea submission and submission policy
Maddak says that if your item is not patented you would need to provide their project manager, Matthew Magnifico with the following information so a non-disclosure form can be sent.

For NDA, send:

Your full name, address, contact email, phone number, and a good time to reach you by telephone.

If the item is patented then contact them by email or phone with details on the idea.

MK Diamond Products

Stuff they sell
Saws, Tools, Accessories and more for Tile, Concrete, Lapidary, Stone, Masonry, Coring, Floor Prep, Fire Rescue

Website
http://www.mkdiamond.com/

HOW TO REACH THEM

Address
MK Diamond Products, Inc.
1315 Storm Parkway
Torrance, CA 90501

Phone Number
(800) 421-5830 (General inquiry)

Email
Frank_Gleason@mkdiamond.com

Some of their products
Tile saws, Floor scrapers, Surface grinders, Cup wheels, Floor scraper blades, Lapidary hand tools, Hand polishing pads and polishing discs for stone, Dry grinding wheels, Core bits for masonry and much more.

Idea submission and submission policy
MK says that the most important consideration is that you have clearly documented and have patented your invention. MK Diamond has a state of the art manufacturing facility in Torrance, CA and a solid distribution network.

If you have a patented product idea and are ready to proceed, contact Frank Gleason.

Mommy's Helper

Stuff they sell
Infant and Toddler items

Website
http://www.mommyshelperinc.com/

HOW TO REACH THEM

Address
Mommy's Helper, Inc.
P.O. Box 780838
Wichita, Kansas 67278-0838

Phone Number
(800) 371-3509
Fax (800) 678-5644

Email
service@mommyshelperinc.com
heather@mommyshelperinc.com

Some of their products
Froggie inflatable tub, Froggie step stool, Corner safe-er-grip, Juice box buddies, Splat mat, Drain n Dry, Bottle Keeper, Sip n snack, Car seat sun shade, Diaper bags, and more

Idea submission and submission policy
Contact Mommy's Helper for a confidentiality agreement and email to:

heather@mommyshelperinc.com, or print and fax document via the above fax number. Once that has been received, you can submit your product idea.

Nike

Stuff they sell
Footwear, Apparel, Accessories, Sports Equipment, Sports Products.

Website
Home page: http://www.nike.com/

HOW TO REACH THEM

Address
Nike, Inc.
Idea Submissions
One Bowerman Drive DF 4
Beaverton, OR 97005

Phone Number
(503) 671-5727

Email
idea.submissions@nike.com

Some of their products
Running shoes, Trail shoes, Performance boots, Tee shirts, Woven tops, Headwear, Hoodies, Duffel bags, Training vests, Power ball, Socks, Sweat bands, Sports tops, Training shorts, Sports bra, Sports bands, Golf balls, Golf bags, Golf shoes, Golf shoe bags, Totes, Golf gloves, Power band, Balance board, Knit

beanie cap, Soccer balls, Shin guards, Swim goggles, Swim suits, Hydro towel, Jackets, Volleyballs, Volleyball knee pads, Ball pump, Footballs, Football gloves, Cleats, Back packs, Tennis skirts, Tennis racquet bag, Running watch, Baseball glove, Water bottles, Baseball bat bag, and much more. See site for product list.

Idea submission and submission policy
Nike asks that you have a patent for your idea. If you already have a patent, please call Nike at (503) 671-5727 for further information.

If you would like to request a copy of the submission guidelines and agreement, send your name and mailing address to email or address listed above.

Please allow 4 weeks for delivery of the guidelines and agreement.

Interview with Gerard Bonner Inventor of the Daddy Pak

Your idea for the Daddy Pak was a result of not wanting to carry the girlie diaper bags on the market as well as wanting a hands free model. How did you come up with the Daddy Pak design you are now using?

Actually I came up with the idea for the Daddy Pak while chasing after my own kids, lugging around my wife's diaper bag and thinking there has to be an easier way than this. I mean there I was the big bad daddy out playing with my kids and I have this floral anchor hanging off my shoulder--didn't look that masculine I got to tell you.

How long did it take you from concept to having a salable product and what were the basic steps you followed?

It took about 2 1/2 years from the time I drew up a schematic then found someone to make a prototype then finally secure a manufacturer that could produce it at a costly venture. Google, google, google. You can find out tons if you put your time in.

Where did you find your daddy product testers?

I "hit the pavement" as they say and asked everyone I saw with a child to try it on and give me their opinion. We also sent out a ton of samples to many blog sites and asked them to "review" it for us and tell us what they thought.

How did you locate a manufacturer and a supplier of the materials used to make the Daddy Pak?

I just kept asking and never gave up. Everyone I talked to, I would ask them what they knew about manufacturers. As it turned out, a good friend of mine happened to have a cousin who did some work with a fellow in China. Turns out this fellow knew half the Eastern seaboard in China and lucky for me knew a manufacturer that happened to make backpacks. After some negotiations we jump started our business and have not looked back. You hear the expression "it's all in who you know" and I can attest it really is true except that, "I didn't know who I didn't know".

What was the most difficult aspect of this process?

I would have to say keeping my patience. I wanted this to take off and be the next biggest thing but there is a reason that some things do not happen the way you think they will. I was not ready for a big fulfillment house at that time or shipping methods, not to mention the salesmanship that I had no experience in or the different process that each chain store employs, price points, lead times, etc. Also now I realized we have started a "brand" instead of just a company, so look out for Daddy Pak wherever you see kids.

Did you have any former experience or education in product invention or was this a, learn as you go adventure for you?

This was definitely a, learn as you go process for me. I am a firefighter in NYC and a finish carpenter on the side (my 2nd job). I didn't know squat from prototypes to manufacturers to dealing with companies in China, but I've come a long way. Persistence, consistency and commitment have led me to a road I was not sure existed, but I am so grateful and thankful that I pursued it and didn't lose sight of my dream no matter what the nay-sayers said.

Is there anything you would have done differently in how you proceeded with your idea? If so…what and why?

Lots. I would have gotten orders before I got product. I would have found publicity connections first. I did try to sell my idea to the other major companies in the dad gear marketplace but got pushed aside, that's when I knew I was onto something. I would have secured my mentors long before I actually did. They have been down the road you want to go long before you and they can direct you where the curves will be, saving you months, money and time, not to mention their expertise.

My mentors have become my good friends, maybe because I have a great personality or maybe because we are from the same mold, but I definitely could not have succeeded if it were not for my mentors. However, at this time I can say that I would truly not change a thing because it has made me and my company who we are today.

Where can the Daddy Pak be purchased?

The Daddy Pak diaper bags for dads can be purchased directly through our website @ www.daddypak.com or at some local stores near you. Please check our website for listings.

One Step Ahead

Stuff they sell
Baby products, including bath items, clothing, safety, travel, nursery, feeding and toys

Website
Home page: http://www.onestepahead.com

HOW TO REACH THEM

Address
One Step Ahead/Leaps and Bounds
P.O. Box 517
Lake Bluff, IL 60044-0517

Phone Number
One Step Ahead: (800) 274-8440
Leaps and Bounds: (800) 477-2189
Fax: (847) 615-2162

Email
Emails can be sent directly from *contact us* page

Some of their products
Learning toys, Fireplace safety guard, Fire escape ladder, Training pants, Bed rail, Kid's pillow covers, Travel potty, Spill proof snack buddy, Sip a bowl, Booster seat tote, Travel bed, Book shelf, Safety scissors, Puzzles,

Twist-trike, Pacifiers, Nursery décor, clothing and much more.

Some of their brands
Mommy's Helper, Summer Infant, One Step Ahead, North American Bear, Learning Resources

Idea submission and submission policy
One step ahead says that they are always on the lookout for new and innovative products for parents, babies and children.

To submit a product for consideration, at *home page*, go to, *contact us* link and click on, *vendor product submission* link. Here you can complete their website submission form.

Once you have uploaded photographs or other documents, you will have the opportunity to provide your contact information. One Step encourages you to provide as much detail as possible. The more information you send the better they can evaluate your product.

A photograph or illustration must be included with your submission. They also accept brochures, price lists or other marketing materials in a PDF or Word document.

Oreck

Stuff they sell
Vacuums, Floor Machines, Small Appliances, Air Purifiers, Accessories and Supplies, Vacuum Bags and Belts, Cleaning Products, Pet Care Solutions

What ideas they will review
New products, marketing methods, finances, improved methods, formulas, processes, designs, artwork, improved machinery, equipment, and improved raw materials and component parts relating to Oreck's line of business.

Website
Home page: http://www.oreck.com/

Idea submission information:
http://www.oreck.com/ ideas/

HOW TO REACH THEM

Address
Oreck Corporation
Intellectual Property- Law Department
Attn: Sr. Paralegal
565 Marriott Drive, Suite 300
Nashville, TN 37214

(Use the above address for all unsolicited submissions)

Phone Number
(800) 989-3535 (Questions or comments)

Some of their products
Oreck Orbiter®, Ultra multi-purpose floor machine, Oreck Rinse-A-Matic steemer Ultra®, Oreck XL Shield Power Scrubber, The Oreck XL® Tabletop air purifier, Cordless ElectrikBroom, Oreck Restaurateur® Floor Sweeper - 12.5, Refrigerator air purifier, Oreck Cordless Zip Vac®, No Return® Pet odor and stain remover. To see full list, visit website.

Idea submission and submission policy
Go to idea submission link listed above or (1) Go to *home page* (2) Go to *contact us* link (3) Click on *submit an idea* to find submission page. Here you can download a Non-Disclosure Agreement (NDA) and Submission Form from submission page. Sign and return as instructed.

This agreement states that no confidential relationship or obligation of secrecy is created between the submitter and the Oreck Corporation by the submission of the idea and its consideration by their company.

All submissions must be made in the form of written descriptions and/or drawings which explains what the idea or suggestion is; how it works, and how it is different and/or better than existing processes or products.

Ortho

Stuff they sell
Indoor Pesticides, Outdoor Pesticides, Weed Killers, Sprayers, Garden and Landscape Disease Control.

Website
Home page: http://www.scotts.com/smg/

HOW TO REACH THEM

Address
The Scotts Company and Subsidiaries
14111 Scottslawn Rd.
Marysville, OH 43041

Phone Number
(888) 270-3714

Email
ortho@scotts.com

Some of their products
Ortho® Weed B Gon MAX® Plus Crabgrass Control Ready-to-Use, Ortho® Weed-B-Gon MAX® Plus Crabgrass Killer Singles™, Ortho® Weed-B-Gon® Chickweed, Clover and Oxalis Killer for Lawns, Ortho® MAX Poison Ivy and Tough Brush Killer Concentrate, Ortho® MAX Premium Pressurized Tank Sprayer, Ortho®

Dial 'N Spray® Multi-Use Hose-End Sprayer, Ortho® MAX® Fire Ant Killer Broadcast Granules, Ortho® Bug-B-Gon® MAX® Lawn & Garden Insect Killer Ready-to-Spray, and much more. See website for complete list.

Some of their brands
Morning Song, Whitney Farms, Hyponex, Super Soil, Miracle Gro, Round Up, Smith & Hawken, and more.

Idea submission and submission policy
Ortho will only accept or consider creative ideas or suggestions related to products or marketing plans that have been patented or are in the public domain (already being sold).

Anything you send that has not been patented or is not in the public domain will be treated as non-confidential and non-proprietary and Ortho will be free to use those ideas or concepts without compensation to you.

Pacific-Cycle, Inc

Stuff they sell
Bicycles, Bicycle Accessories

Website
Home page: http://www.pacific-cycle.com/

HOW TO REACH THEM

Address
Pacific Cycle, Inc
Idea Submission
4902 Hammersley Road
Madison, WI 53711

Phone Number
Phone: (800) 626-2811
Fax: (800) 858-2800

Email
info@pacific-cycle.com

Some of their products
Mountain bikes, City/Commuter bikes, Cruisers, Electric bikes, BMX, Fixed wheel and swivel wheel strollers, Bike trailers, Pedal cars, Jogging stroller infant headrest, Bicycle trailer storage pouch, Stroller weather shield, Jogging stroller computer, and more.

Some of their brands
Mongoose, Schwinn, Roadmaster, Pacific, Instep, KidTrax, Pacific Outdoors, GT

Idea submission and submission policy
Contact Pacific Cycle at the above email address or the listed street address for Idea Submission and NDA forms. Submissions should be made via regular mail at above listed address.

Pelham West Associates

What they do
Pelham West is a product scout company. They identify and evaluate new products and inventions from independent inventors seeking to match those products with interested businesses through licensing of product patents, or their patent potential.

Website
Home page: http://www.pelhamwest.com/

Contact page/submission link:
http://www.pelhamwest.com/Contact_Us.htm

HOW TO REACH THEM

Address
N/A

Phone Number
N/A

Email
info@pelhamwest.com

Stuff they are looking for
Medical Products used in surgery such as new surgical staplers, etc. Only products with several metal components such as small stamped parts, stamped parts with some

machining and other coiled springs or wire forms. Possibly an assembly with plastic components along with the stamped metal parts and springs.

Simple Security Products: Simple gadgets or products that help secure doors, windows, etc. by adding to or supplementing existing locks, knobs, hinges, etc. No electronics or alarms.

Storage Products or Systems: Any organizational and/or storage products for offices, basements, homes, garages, etc.

Cord Related Products used for storage, safety, or organizing and keeping wires and cords neat to be sold commercially or in office supply stores or home centers.

Handy Gadgets, Hangers, etc. that are simply made and can be used for organizing closets, tool sheds, basements, etc. Very interested in simple to manufacture products, especially bent wire or other brackets, hangers, etc. that can be manufactured in the U.S. and brought to market quickly

Safety Products for Industry used to make factories and workplaces safer, particularly in the area of cord and cable management. Products that can reduce trip hazards and products used to reduce insurance liability in business, office settings or tradeshow and meeting venues.

TV or Catalog Products in house wares, kitchen, bed/bath, tools, auto, personal care. These products must solve a problem and be demonstrable on TV or lend themselves to catalog sales. Non-seasonal products that can be sold throughout the United States are preferred.

Outdoor Games and Toys. Only games that are played outside will be reviewed, including toys or games for use in swimming pools and water play.

Emergency Products or Tools used in response to natural disasters or other emergency situations, including water rescue products, ice rescue products, lifeguard equipment, products that sell to fire and rescue departments, the Coast Guard, FEMA, etc. or other government agencies, as well as products that can be used to prevent or mitigate damages from flooding or other natural disasters.

Products that Help Relieve or Prevent Back or Joint Pain or that can provide ergonomic benefits, such as new seating, sleeping, or computer accessory products. (Examples: ergonomic folding furniture, exercise or massage equipment). No herbs, oils, medicines or vitamins.

Office Products sold in retailers such as Office Max, Office Depot or Staples, including new packaging concepts.

Automotive Accessories, especially winter-related products such as ice scrapers, folding shovels, etc.

Housewares or Gadgets, especially items for the kitchen sold in kitchen shops or retail stores such as Bed, Bath and Beyond or Linens 'N Things. Top priority given to under $5.00 retail products that can be sold on clipstrips. Also simple products that help with food preparation.

Gadgets, Garden Accessories, etc. that sell in Home improvement stores. Products that allow the aging population to continue to enjoy gardening.

Products with Multiple Versions for Homeowners, Contractors or Government Use.

Cleaning or Personal Hygiene Products, in particular, items for the kitchen or bath with a consumable component. New dispensing systems for cleaning, hygiene products, scents and air fresheners.

Maintenance Supplies and Accessories*:*
Items for Departments of Public Works, state or federal highway departments, military bases, national parks, etc. for repair and maintenance of property and equipment.

Idea submission and submission policy
Go to submission link listed above or (1) Go to *home page* (2) Click on *Inventors Go Here* (3) Go to *products wanted.*

After reviewing their needs, if you desire to submit an idea, click on the link, *contact us for product submission,* located at the bottom of the page and follow instructions.

Thoroughly read the information in all of the links to get a full understanding of what Pelham West does and if your product suits their needs.

Comments
There is no out of pocket fee for the inventor. Pelham West is paid by client companies rather than by inventors.

For a small fee, Pelham also offers consulting services to determine the best commercialization opportunities for your invention. To set up a telephone appointment, send an email that includes the following:

1. A brief non-confidential description of your product.

2. The questions you would like to have addressed.

3. A list of the times you are available.

Pfizer

Stuff they sell
Products for Animal and Human Health

Website
Home page: http://www.pfizer.com/home/

Submissions page:
http://www.pfizer.com/research/licensing/conta
ct_us.jsp

HOW TO REACH THEM

Address
Pfizer Inc
235 East 42nd Street
New York, NY 10017
USA
(This address is not for submissions)

Phone Number
(212) 733-2323

Some of their products
Ben Gay, Efferdent denture cleanser tabs,
Listerine, Lubriderm lotion, and much more.

Idea submission and submission policy
Any technology, device, diagnostic, animal health or human health business opportunity at any stage of development may be submitted.

To Submit: Log onto http://www.pfizer.com/partnering, enter your contact and product information under *submit your idea* link. You can also access this via the submission link above.

Comments
Current areas of interest include, but are not limited to, Alzheimer's disease, diabetes, inflammation and immunology, oncology, pain, psychoses, vaccines, pharmaceutical science, asthma, infectious disease, COPD.

Poof-Slinky, Inc.

Stuff they sell

Foam Sports Balls, Action Toys, Mini Sports Balls, Flying Toys, Slinky®, Novelty Toys, Ideal® Classics, Activity Books, Fun Labs, Activity Kits, Mini-Labs, Science Kits, Magnets, Tee Pees, and more.

Website

Home page: www.poof-slinky.com

HOW TO REACH THEM

Address

Poof-Slinky, Inc
45400 Helm St.
P.O. Box 701394
Plymouth, MI, 48170-0964

Phone Number

(734) 454-9552
Toll free: (800) 329-TOYS
Fax: (734) 454-9540

Email

jbrady@poof-slinky.com
sales @poof-slinky.com
kmarcum@poof-slinky.com

Some of their products
Foam soccer ball, Foam volleyball, Dart Launcher, MiniMega Bow, Gold plated Slinky®, Slinky® batons and pom-poms, Back Pack Party Set, Mr.Machine, Rack 'N Roll Bowl, Sure Shot Hockey, Sure Shot Hover Blaster, Dot to Dot books, Kiddie Color books, Fingerprint Files, Power Putty Slimes, Secret Codes, Frontier logs, Fabulous Fiddlestix®, and much more. See website for more products.

Some of their brands
Ideal®, Slinky®, Poof®, Buki®, and more.

Idea submission and submission policy
From the above contact information, please request and review Poof-Slinky's Non-Disclosure Agreement. Sign and return with your product submission to John Brady, VP of operations. Product concepts are reviewed and you will be notified accordingly.

Pittsburgh Corning Corporation

Stuff they sell
DecoBloc (Specially designed glass block for artists and crafters to showcase creative expression), Grill Brick (Cleans grill surfaces), Glass block for buildings, Foamglas® (cellular glass insulation)

Website
Main home page:
https://customers.pghcorning.com
http://www.decobloc.com/home.asp
http://www.foamglasinsulation.com/
http://www.grill-brick.com/
http://www.pittsburghcorning.com

HOW TO REACH THEM

Address
Pittsburgh Corning, Corp.
800 Presque Isle Drive
Pittsburgh PA 15239

Phone Number
(800) 545-5001

Email
Elaine_Underwood@pghcorning.com

Some of their products
Glass block, Mortar applications, Kwik n EZ® Products, PC Glass block sealant, Provantage® Applications, Foamsil® Insulated travel mug and other Foamsil® products, DecoBloc, Grill Brick

Idea submission and submission policy
If you would like Pittsburgh Corning to consider your idea related to glass block, please submit it in the following manner:

1) If your idea is patented or protected by trademark or copyright laws, please submit this information with your idea, along with a copy of the patent, trademark or copyright registration certificate.

2) If your idea is not subject to patent, trademark or copyright protection, PCC requires a waiver before considering your idea. Request a waiver and if you agree with the terms and conditions set forth, please sign the form and return it with a written description of your idea.

Once they have reviewed your information, they will contact you with a response. Please return all information to the attention of *Elaine Underwood.*

Comments
Go to *Pittsburgh Corning* heading at "customers" website above to view links to PCC's product lines.

Plaid

Stuff they make and sell
Bucilla Needle Crafts, Craft Painting, Decorative Painting, Decoupage, Fabric Decorating, Glass Painting, Home Decorating, Jewelry Making, Kids Crafts, Knit-Wit & Doodle-Loom, Mosaics, One Stroke Painting, Scrapbooking, Stenciling & Stamping, Rubber Stamping & Card Making

Website
Home page: http://www.plaidonline.com/

Idea Submission page:
http://www.plaidonline.com/ideas

HOW TO REACH THEM

Address
Plaid Enterprises, Inc.
P.O. Box 7600, Norcross
GA 30091-7600 (Not for submissions)

Phone Number
(800) 842-4197 (Consumer Advisory Team)

Some of their products
Paints, Wood surfaces, Clear stamps, Brass stencils, Foam mounted stamps, Cross-stitch, Needlepoint, Crewel, Plaster, Molds, Books, Enamels, Fabric paints, Pigments, Stained

glass (look) leading, Window color, Knitting tools, Mosaic tiles, Mosaic embellishments, and much more. See website to learn more.

Some of their brands
Apple Barrel®, All Night Media®, Bucilla®, Cork Stamps, Faster Plaster®, Folk Art®, Gallery Glass®, Instant Expressions®, Knit-Wit®, Doodle Loom®, Make it Mosaics®, Mod Podge®, One Stroke ™ Painting, Simply Stencils® & Stamps, Stencil Décor®, Connect™, Artistrywear™

Idea submission and submission policy
Go to idea submission link above or (1) Go to *home page* (2) Click on *idea submission* link at bottom to submit idea.

Plaid is currently accepting submissions for new product ideas and innovative crafting techniques.

Comments
All submissions are to be made via Plaid's submission web page. Listed phone number and address are not for submissions.

Powermate

Stuff they sell
Air Tools, Generators. Other.

Website
Home page: http://www.powermate.com/

HOW TO REACH THEM

Address
Invention Submissions Powermate Corporation
4970 Airport Rd.
P.O. Box 6001
Kearney, NE 68848

Phone Number
(800) 445-1805

Email
N/A

Some of their products
Pressure washers, Air compressors, Portable Generators, Air tools, Cordless soldering tool, Firelight propane fireplace, Powerworks 225 watt inverter, Sport Cat Perfect Temp catalytic heater, Air tools, Grinders, Drills, Ratchets, Nailers, Staplers, Impact wrenches, Spray guns, and much more.

Some of their brands
Black Max, Industrial Air, Magna Force, Powermate, Coleman Powermate, Endura, ProForce, Progen, Sanborn and private labels that include Durabuilt, Husky, Marshalltown, Husky and Kobalt.

Idea submission and submission policy
Obtain a submissions policy by calling (800) 445-1805, or writing to the above address.

Comments
Make sure that your ideas are submitted in accordance with Powermate's Corporate policy or they will be returned, or destroyed without being reviewed.

Pradco

Stuff they sell
Fishing lures

Website
Home page: http://www.lurenet.com/

HOW TO REACH THEM

Address
For freshwater products, please send your packet to:
PRADCO - Fishing
Attn: Jason Needham
3601 Jenny Lind
Fort Smith, AR 72901

For saltwater products, please send your packet to:
PRADCO - Fishing
Attn: Kim Norton
3601 Jenny Lind
Fort Smith, AR 72901

(Addresses are for product idea submission)

Phone Number
(479) 782-8971

Email
N/A

Some of their products
Fishing Lures, Game Calls, Scents, Attractants, Game Decoys

Some of their brands
Excalibur®, Rebel®, Bomber®, Cotton Cordell®, Knight Hale®, Carry-Lite®, Code Blue®

Idea submission and submission policy
Please send a sample, photo or drawing of your product that you would like to introduce with as much information as you can to Pradco. Please include any patent or trademark information.

Pradco does not sign non-disclosure forms. If they decide to go forward with your product line they will sign a contract at that time.

Presto

Stuff they sell
Small Appliances, Cookware

Website
Home page: http://www.gopresto.com/

Submission info page:
http://www.gopresto.com/information/inventor

HOW TO REACH THEM

Address
N/A

Phone Number
N/A

Email
inventions@gopresto.com

Some of their products
Knives, Cutting boards, Pizza ovens, Bacon cookers, Canners, Popcorn poppers, Frying pans, Electric peelers, Knife sharpeners, Griddles, Heaters, Multi-cookers, Shoe polishers, Shredders, Waffle makers, and more.

Idea submission and submission policy
Go to submission page (see link above) or (1) Go to *home page* (2) Click on *inventor info* at bottom of the page and email a description and/or picture(s) of your idea along with contact information to:

inventions@gopresto.com

If you do not receive a response from National Presto within 30 days of your submission, assume that National Presto will not be pursuing your idea.

If they are interested in your product idea you will be informed either by email or telephone.

Comments
Send any submission questions to the above email address.

Proctor & Gamble

Stuff they sell

Air Fresheners, Antiperspirants, Deodorants, Hair Care, Hair Color, Healthcare, Household Cleaners, Laundry & Fabric Care, Colognes, Baby & Child Care, Body Wash and Soap, Batteries, Colognes, Commercial Products, Cosmetics, Prescription drugs, Pet Nutrition, Shaving, Prestige Fragrances, Skin Care, Dishwashing, Oral Care, Feminine Care, Paper Products, Small Appliances, Snacks & Coffee.

Website

Submission home page:
www.pgconnectdevelop.com

P & G Home page:
http://www.pg.com/en_US/index.shtml

HOW TO REACH THEM

Address

N/A

Phone Number

N/A

Email

Email questions can be sent through the *contact us* link on the website.

Some of their products
Toothpaste, Laundry soap, Dryer sheets, Fabric softener, Shampoo, Diapers, Face wash, Cough syrups, Dog food, Laxatives, Potato chips, Coffee, Razors, Make-up, toilet paper, and much more. See *product index* at website for product line.

Some of their brands
Gleem®, Bounce®, Olay®, Vicks®, Head & Shoulders®, Cover Girl®, Herbal Essence®, Home Café,® Always®, Camay®, Crest®, Gillete Fusion Power®, Downey®, Luvs®, Pepto Bismal®, Febrese®, Old Spice®, Puffs®, Pringles®, Noxema®, Swiffer®, and more. See website for more of P & G's brands.

What they are looking for
P & G is looking for new products, technology and business models or methods. A list of more specific needs that P&G is currently looking for is found under the Browse P&G's needs link.

Idea submission and submission policy
Go to P & G's submission *home page* link above and browse the list of needs via their link. If your innovation suits P & G and meets the criteria listed below, register and go to the submission link to submit your idea.

P&G will not accept any idea for consideration unless it is protected by a patent or a patent application. Submissions can only be made via P & G's website.

Comments
P & G's site says that it might be interested in your idea if it meets one or more of the following criteria:

1. Your innovation addresses a large, unmet consumer need.

2. Your innovation offers something new to an existing P&G category or brand.

3. Your packaging solution has been demonstrated.

4. Your technology is proven and can be quickly applied to a P&G consumer need.

5. You have a game-changing technology or approach.

Purefishing

Stuff they sell
Sporting and Athletic Goods

Website
Home page: http://www.purefishing.com/

HOW TO REACH THEM

Address
Pure Fishing Corp.
7 Science Court
Columbia, SC 29203

Phone Number
Toll Free: (800) 334-9105

Email
Contact Jessica Woods at:
jmwoods@purefishing.com (idea submission)

Some of their products
Fishing equipment; Sporting goods, Stainless steel shafting, Steel shafting, Fishing reels, Fishing lures, Rod, Fishing bait, Fishing weights or sinkers, Fishing tackle, Foil, Fishing rods, Grating, Fishing line, Profiles, Strip, Piling, Shafting, Rails, Billets, Ingots, Honeycomb core.

Idea submission and submission policy
Using the above email address, request a Non-Disclosure Form from Pure Fishing and include your contact information. The NDA must be filled out, signed and returned before they will receive or review your product idea.

Prym Consumer USA, Inc

Stuff they sell
Sewing, Quilting, Home Décor, Fashion, Needlecraft

Website
Home page: http://dritz.com/

Submission page info:
http://dritz.com/askus/submissions/

HOW TO REACH THEM

Address
Product Submissions
Prym Consumer USA, Inc. Corporation
P.O. Box 5028
Spartanburg, SC 29304

Phone Number
N/A

Email
productsubmissions@dritz.com (Submission questions)

Some of their products
Pins, Needles, Marking pens for quilting, Cutting mats, Cutters, Scissors, Tools and sewing accessories for home decorating

projects, Line of purse handles, Button on bags, Fabric dyes, Iron on designs, Cross-stitch, Needlepoint, Beading, Sewing notions, and more.

Some of their brands
Dritz, Collins, Sewing Basket, Prym Sewing, Kimberly Poloson, Dritz Cutting, Quilting Basket, Fons & Porter, Dritz Home, Bag Boutique, Dylon, Painted Treasures, Omnigrid, Fashion Embellishment, Boutique, LoRan, Iron On Letters, St. Jane, Mary Engelbreit.

Idea submission and Submission Policy
Go to submission link above or (1) Go to *home page*, (2) Link to *ask us*, (3) Click on *product idea submissions*.

Prym has a four step plan:

1. You must send Prym your contact information either online or via snail mail.

2. Complete and return standard confidentiality agreement.

3. Product concept submission.

4. Product management review.

Their Product Management Team meets every 4-8 weeks.

For any questions, including the dates of the next review meeting you may email them at any time at: productsubmissions@dritz.com

Quantum

Stuff they make and sell
Fresh water and Salt water Rods and Reels.

Website
Home page: http://www.quantumfishing.com

HOW TO REACH THEM

Address
Quantum
6105 E. Apache
Tulsa, OK 74115

Phone Number
(800) 588-9030 (Questions: Customer Relations)

Email
email.quantum@zebco.com (Customer Service)

Some of their products:

Saltwater Conventional Reels: Blue Runner, Aruba PT's.

Saltwater Bait Cast Reels**:** Energy PT's, Iron, Cabo PT's, Accurist PT's.

Saltwater Spinning Reels: Boca PT's, Surge, Catalyst PT's inshore.

Saltwater Rods: Cabo PT's (Inshore and Offshore), Blue Runner (Inshore, Offshore, and Solid Glass).

Freshwater Bait Cast Reels: Tour Edition PT, Energy PT, Mantra, Icon.

Freshwater Spinning Reels: Tour Edition PTi, Energy PTi, Kinetic Pti.

Freshwater Rods: Harmonic, Big Cat, Cold Water, Teton Trout, Affinity. AVS(Advanced Youth System), and much more.

Idea submission and submission policy
Quantum will only accept submissions which have a valid patent, and where the patent owner is submitting the idea. Patent pending ideas will not be considered. Write the patent number on the outside of the submission package. No email submissions.

Any unsolicited letter or package that is sent to Quantum that is identifiable as an unpatented idea will be discarded unopened. All others will be discarded as well immediately upon identification as an unsolicited idea. All e-mails will be immediately deleted. Quantum can make no assurances that your ideas and materials will be treated as confidential or proprietary.

QVC

What they do
QVC is a multimedia retailer

Stuff they sell
Jewelry, Fashion, Beauty, Cooking, Dining, Home Products, Electronics, and Sports & Leisure.

Website
Home page: http://www.qvc.com/

HOW TO REACH THEM

Address
QVC
1200 Wilson Drive
Mail Code: 128
West Chester, PA 19380
(Send requested samples only)

Email
Vendor_Relations@QVC.com

(If you still have questions after viewing web information, contact QVC via this email address)

Some of the products they sell
Handbags, Blazers, Dresses, Skirts,

Sleepwear, Earrings, Necklaces, Ankle Bracelets, Cosmetics, Grills, Kitchen electrics, Seafood, Art, Rugs, Mattresses, Furniture, Collectibles, Books, Movies, Music, Cameras, Camcorders, and much more. See website for full product list.

Product submissions
Go to *home page* and (1) Click on *Vendor Relations* link (2) Click on *How to become a QVC Vendor* in the side column (3) Go to *Begin the submission process* at the bottom of page.

If QVC requests a product sample send it to the above address.

Comments
Your products must already be manufactured.

Reckitt Benckiser

Stuff they sell
Health & Personal, Fabric Care, Household Care.

Website
Home page: http://www.reckittbenckiser.com

Submission/innovation page: http://www.rb-idealink.com/

HOW TO REACH THEM

Address and phone number
The following will link you to a number of worldwide contacts for Reckitt Benckiser depending on your location:

http://www.rb.com/RB-worldwide/Contacts

Email
N/A

Some of their products
Dishwasher detergents, Face wash and cleansers, Stain remover, Disinfectants, and much more.

Some of their brands
Veet, Lysol, Calgonit, Electrasol, Woolite,

Spray ‘n Wash, Vanish, Clearasil and more. See website for complete list.

Idea submission and submission policy
RB is looking for a wide variety of innovative ideas. If you have a product already on the market or close to market ready, they would like to know more about it.

Go to submission/innovation website link listed above or (1) Go to *Home page* (2) Click on *Innovators* (3) Go to *Got a great idea/invention* link (4) Go to *How to Submit* link.

Submit a technology or a product via the RB website. RB will take approximately 3 months to review your submission and if necessary will then request further information, samples or prototypes.

Comments
Although RB will review other products, they have their own “wish list”. View RB’s *Most Wanted* for current product ideas they are in the market for.

Rubbermaid

Stuff they sell
Rubbermaid: Bath, Beverage, Garage, Shelving, Closet, Kitchen, Storage, Coolers, Laundry, Utility, Food Storage, Outdoor, Waste & Trash. Rubbermaid Builder: Wood Closets, Wire Closets, Garage Systems.

Website
Home page: www.newellrubbermaid.com

Note: Newell Rubbermaid no longer accepts unsolicited ideas or ideas from individuals, but will review those submitted by a business entity if it falls under the category of technological items they have posted.

HOW TO REACH THEM

Address
N/A for submissions

Phone Number
N/A for submissions

Email
N/A. For general information questions use *contact us* form at website.

Some of their products
Closet organizers, Coolers, Storage units, Bike hooks, Hose hook, Wheelbarrow holder, Food storage containers, Storage shed, Shelving, Trash cans, Hampers, Laundry baskets, Vacuums, Cleaning carts, Utility carts, Scarf and tie racks, Shoe cubbies, Wire baskets, Built in hampers, Jewelry drawers.

Some of their brands
DuraChill™, Produce Saver™, Easy Find Lids™, Configurations™, FastTrack®, TightMesh®

What they are presently looking for
Patented game-changing technology that addresses unmet consumer needs. Patented or published new materials, ground-breaking manufacturing processes or novel ways of doing business that can be used to innovate and make products faster, better or at a lower cost.

Idea submission and submission policy
(1) Go to home page (2) go to *Our Company* (3) Click on the link, *Do Business with Us (4)* Go to *technology assets* link for information on current technological interests.

Rubbermaid's requests are always focused on a particular technology, product category, or consumer need and are not open calls for unsolicited ideas. When responding to a specific area of interest, your idea must meet several requirements. For instance, your idea must be patented or patent pending with a published patent application.

Detailed information about how to respond to a specific Newell Rubbermaid request for information or innovation challenge is provided on their website.

Responses will only be reviewed if they are submitted through this website, and if they meet the requirements of the specific request.

Safety 1st

Stuff they sell
Baby Products

Website
Home page: http://www.safety1st.com/

Submissions page:
http://www.safety1st.com/great_idea/

HOW TO REACH THEM

Address
Dorel Juvenile Group
Consumer Relations Department
P.O. Box 2609
Columbus, IN 47202-2609

Phone Number
(800) 544-1108
Fax: (800) 207-8182

Email
Go to *contact us* link for email contact.

Some of their products
Monitors, Car seats, Strollers, Walkers, Gates, Booster seats, Bath tubs, Potties, Bouncers, Swings, Thermometers, Bath accessories,

High chairs, Toys, Bed rails, Diaper pails and more.

Idea submission and submission policy
Go to submissions page link listed above or (1) Go to *home page* (2) Click on *Got a Great Idea* and download their Mutual Non-Disclosure Agreement.

Enter your name in the signature box, save the file and upload at the bottom of the website form.

Comments
Presently, Safety-First does not accept non-patented products.

Scheewe

Stuff they sell
Decorative painting lesson/instruction books, DVD's and Supplies

What they are looking for
Painting projects/lessons for publication in books.

Website
http://www.painting-books.com/

HOW TO REACH THEM

Address
Scheewe Publications
P.O. Box 20518
Portland, Ore 97294

Phone Number
(503) 254-9100
Fax: (503) 252 9508.

Email
Website email form available

Some of their products
Books on oils, Acrylics, Pen & inks, Colored pencil, Watercolor, DVD art workshops, Artist paper, paint sets, and more.

Idea submission and submission policy

Scheewe says: When submitting potential book material, please include the following:

- A minimum of 15 photographs of your artwork. Since they can't see your original art, your pictures should be clear. Professional photos are not required. The more photos you send the better Please do not send any artwork; photos only.

- Include one lesson plan for one of your painted projects. This would include a description of the palette, brushes, supplies, instructions, preparation, base coating, detailing and finishing necessary to complete the project.

- Include one pattern from your projects. Please use black ink on white paper.

- Tell them about yourself. A brief history of your painting background.

Consider the following when submitting material for a potential book:

- All of your designs must be original. Do not send photos from projects you have painted in another teacher's class. The projects you submit must be your own design. Due to copyright laws, they do not publish any material that has been copyrighted previously.

- Please do not send material that has already been printed and sold as a packet. They would be glad to look at any packet in reference to the type of work you do, however, they do not print any material that has been reproduced or sold.

If you have any questions, you may call, write, or e-mail them. Make a copy of your work before you send it. If you would like your material returned, please include a S.A.S.E.

Shuffle Master, Inc.

Stuff they sell
Table Games

Website
Home page: http://www.shufflemaster.com

Submission page:
http://www.shufflemaster.com/04_players/submit_Your_game_idea/contact_info/gameideas/

HOW TO REACH THEM

Address
Russel Chumas
c/o Shuffle Master Inc.
1106 Palms Airport Drive
Las Vegas, NV 89119 USA

Phone Number
Phone: (702) 897-7150
Fax: (702) 897-2284

Email
shfl@shufflemaster.com

Some of their products
Ultimate Texas Hold 'Em, Bad Beat Texas Hold 'Em, and more.

Idea submission and submission policy
Go to *For Players* link and click on, *submit your game idea* or go directly to submission link above.

Fill out and send the online form and Shufflemaster will send you a downloadable idea Submission Waiver Form.

You will need to submit **two** copies of this form via regular mail to the above address prior to any discussions on your submission.

Comments
The above contact information is for those in the USA. For contact information in other countries, go to *contact us* link. (http://www.shufflemaster.com/01_company/contact_us/)

Simpson Strong Tie

Stuff they sell
Construction & Engineering Products

Website
Home Page: http://www.strongtie.com/

Submission Page:
http://www.strongtie.com/ideas/index.htm?source=topnav

HOW TO REACH THEM

Address
Simpson Strong-Tie Company, Inc.
5956 W. Las Positas Boulevard
Pleasanton, CA 94588

Phone Number
(925) 560-9000
Fax: (925) 847-1603

Some of their products
Cold Form Steel: Anchors, Concrete Connectors, Hangers, Masonry Connectors, Shearwall, Truss Connectors.

Concrete Connectors: Bearing Plates, Anchor Adhesives, Anchor Bolts, Beam Seats, Bearing Plates, Form Ties.

Engineered Wood and Structural Composite Lumber Connectors:
Adjustable Hangers, Concealed Joist Tie, Hip Connectors, Purlin Anchors.

For complete list, go to: *Browse Product* link. (http://www.strongtie.com/products/category_list.html)

Idea submission and submission policy
On *Home Page*, (1) Go to *Customer Service* link (2) Click on *New Product Request*.

This will take you to the submission page. Choose the appropriate link depending on whether your invention falls under patented, patent pending or provisional patent.

Here you will need to print a submission form and send to the above address, or fax completed form to the above fax number to the attention of, *Tawn Simons.*

Include any samples, drawings or photos that may help with the description of your invention.

Comments
Must have a patent, provisional patent, or patent pending.

Please read all forms included in the PDF file link. Forms include New Product Request Form, Submission Policy and Disclosure Form and Submission Policy and Disclosures.

Submissions missing any or all of the forms will not be considered. Simpson Strong-Tie will not sign confidentiality agreements.

Spin Master

Stuff they sell
Toys

Website
Home page: http://www.spinmaster.com/

HOW TO REACH THEM

Address
Spin Master Ltd.
450 Front Street West
Toronto, ON
M5V 1B6 Canada
(Not for submissions)

Phone Number
(800) 622-8339

Email
Website form

Some of their products
Air Hogs, Girl Crush Tattoo Maker, Hershey's Chocolate Maker, Wiggles Guitar.

Idea submission and submission policy
Spin Master is always on the hunt for new inventors, new product ideas. If you are new to inventing and would like to submit a concept

with Spin Master, go to their website *home page* listed above, click on *Contact Us* and go to, *Inventor Relations* to fill out their online submission form.

A few of their questions will include, how far along your innovation is, if it is patented and if it has been submitted to other companies.

Their form does not constitute a formal submission. It is used as a means to determine whether your concept is in a category they are pursuing.

Stamina Products

What they sell
Exercise & Fitness products

Website
Home page: http://www.staminaproducts.com/

HOW TO REACH THEM

Address
Invention Submission
c/o Stamina Design Dept.
P.O. Box 1071
Springfield, MO 65801-1071

Phone Number
(417) 889-7011
FAX: (417) 889-8064

Email
inventors@staminaproducts.com

Some of their products
Kettleballs, Ellipticals, Treadmills, Pilates products, Ab hyper bench, Home gym, Power towers, Videos, Spacemate folding stepper, Electronic stepper, Golf fitness training and more.

Idea submission and submission policy
Stamina Products, Inc. is always seeking breakthrough designs and ideas from independent designers and engineers in the exercise and fitness industry.

Contact them at any of the above means to see if your idea meets their needs.

Stampin' Up!

Stuff they sell
Rubber stamps, Embellishments, Paper, Ink Cetera, Stampin' Memories, Definitely Decorative, Tools, Storage

Website
Home page: http://www.stampinup.com/

HOW TO REACH THEM

Address
Stampin' Up!
c/o Brent Steel
12907 South 3600 West
Riverton, UT 84065

(Submission address)

Phone Number
(800) 782-6787

Email
demos@stampinup.com (Contact them with any questions)

Some of their products
Designer series paper, Envelopes, Eyelets, Ribbons, Rub ons, Brads and Buttons, Accents and Elements, Stampin' pads, Stampin' ink,

Stampin' write markers, Pens and Journalers, Water coloring, Pastels, Embossing plates, Punches, and much more.

Idea submission and submission policy
Stampin Up! says that the vast majority of their artwork is created through their in-house design team. On occasion, they may discover an illustrative style that they cannot execute to their satisfaction. In these situations they will use freelance talent to fill that creative niche.

If you are interested in being commissioned for your artwork, please submit six samples of your work for review by their concept art department, (see address above.) These samples will not be returned.

Whenever an art portfolio is received they try to send an immediate response so the sender knows that their artwork has been received and reviewed. The letter will also indicate whether or not there is a current need for the illustrative style/styles being submitted. Should they decide to commission your work or require additional information they will contact you.

Compensation for commissioned artwork is currently paid in a lump sum for all freelance art (compensation is negotiated at time of contract). All commissioned artwork becomes the property of Stampin' Up! It is not Stampin' Up!'s practice to pay royalties on any artwork.

Staples

Stuff they sell
Office Supplies, Technology, Furniture, Medical Supplies

Website
Submission page/Home page:
http://www.pdgevaluations.com/

HOW TO REACH THEM

Address
N/A

Phone Number
N/A

Email
questions@pdgevaluations.com (If you have any questions)

Some of their products
Binders, Boards and Easels, Computer Bags and Cases, Labels, Label makers, Staplers, Punches, Trimmers, Tape, Rubber bands, Drafting and Art supplies, Calendars, Planners, Computers, PDA's, GPS, Cameras, Calculators, Scanners, Office Machines, Frames, Shredders, Telephones/ Communication, Software, Carts, Stands,

Bookcases, Armoires, Chairs, Cubicles, Desks, Filing cabinets, Lamps, Lighting, Safes, Tables, Chair mats. For a complete list, view Staples website http://www.staples.com

Idea submission and submission policy
Go to submission page listed above and follow instructions. Using the online submission form, state the problem your product idea addresses, what it is that your product does, but *not* how it does it, along with your product's stage of development.

If Staples is interested, you will be sent a Non-Disclosure Agreement (NDA) to fill out and sign before proceeding to the next step. Only the creator of the product/ idea can make a submission. You will be notified within 8-12 weeks.

Comments
PDG, is a third party product screener that receives, compiles, and reviews submissions to ensure that only non-confidential information is evaluated by Staples. Submissions will only be accepted through this website.

Sterilite Corporation

Stuff they sell
Household, Storage, Hardware and Kitchenware

Website
Home page: http://www.sterilite.com

Submission information found @ FAQ page: http://www.sterilite.com/faq.html

HOW TO REACH THEM

Address
Sterilite Corporation
Attn: Idea Submission
30 Scales Lane
P.O. Box 524
Townsend, MA 01469-0524

Phone Number
N/A

Email
N/A

Some of their products
Laundry baskets, Hampers, Wastebaskets, Pails, Caddies, Totes, Under bed storage, Drawer carts, School boxes, Storage crates,

Cabinets, Shelving, Tool racks, Shelf totes, Sink sets, dishpans, Bowls, Pitchers, Ice cube trays, Food storage.

Idea submission and submission policy
Go to FAQ page listed above and click on question: *Does Sterilite accept product ideas from outside inventors?* This will answer submission questions and link you to an online idea submission form for you to print out.

You will need to send Sterilite:

- A signed copy of the submission form.
- A copy of your patent or patent pending documentation.
- Up to three 8 ½ x 11 maximum sized images.

Ideas and inventions submitted that do not include a signed Idea Submission Form will be returned. Submission must also have a valid US patent. Do not submit models, or prototypes.

Stride Tool

Stuff they sell
Tools

Website
Home Page: http://stridetool.com/

HOW TO REACH THEM

Address
Director, Marketing Services
Stride Tool Inc.
30333 Emerald Valley Parkway
Glenwillow, OH 44139

Phone Number
Corporate Offices: (440) 247-4600
Customer Service: (888) 467-8665
Fax: (800) 527-6383

Email
info@stridetool.com

Some of their products
Tube Working Tools: Tube Cutters, Tube Benders, Flaring and Swaging, Tubing Tool Kits, Accessories and Wrenches.

Refrigeration & Air Conditioning Products: Manifolds, Gauges, Hoses, Couplers,

Accessories, Tools.

Electrical & Data Collection Tools: Romex® Tools, Data Communication, Strippers & Crimpers.

Snap Ring Pliers: Convertible, Fixed Tip, Replaceable Tip, Replacement Tips, Lock Ring.

***Wire Twisting Pliers*:** Reversible, Wire Twisters.

Automotive Specialty Tools: Oil Changing, Hones and Deglazers, Battery & Engine, General Auto Tools.

What they are looking for
Tools that are useful to any type of professional tool user, anyone who makes their living with tools, such as an automotive technician, electrician, airplane mechanic, or an industrial maintenance person, etc.

Idea submission and submission policy
On *Home Page*, go to *Inventors* link. Download Submission Form at link, fill out and send to the above address. This must be sent before they will review your product idea.

Stringing Magazine

Who they are
Stringing magazine is a quarterly publication of beaded jewelry and the step by step project instructions to make necklaces, bracelets and earrings.

Website
Home page:
http://www.stringingmagazine.com/

What they are looking for
Beaded/strung jewelry project instructions, more specifically necklaces, bracelets and earrings.

HOW TO REACH THEM

Address
Stringing
Attn: Danielle Fox, Editor
201 East Fourth St.
Loveland, CO 80537-5655

Phone Number
(800) 272-2193

Email
stringingsubmissions@interweave.com

Submission policy
Stringing does not accept submissions that have been previously published. After reading complete submission guidelines at website, send a legible list of the contents of your submission package and your contact information. Include full name, address, phone number and e-mail address.

Also send a text document: An electronic version sent to Stringing via e-mail at stringingsubmissions@interweave.com that includes the following information:

1. Instructions (for one-page projects only)
2. Resource information
3. Bio and photograph of yourself

To access guidelines, go to *Stringing's* website and click on the *contributor's* link.

Comments
Submissions must represent original work.

SwimWays

Stuff they Sell
Toys, Swim Sports, Décor, Lounges and Accessories for the pool

Website
Home page: http://www.swimways.com/

HOW TO REACH THEM

Address
I have the next, great swimways product!
Swimways Corp.
5816 Ward Court
Virginia Beach, VA 23452

Phone Number
(757) 460-1156
Toll Free:(800) 889-SWIM (7946)

Email
inventorsubmissions@swimways.com

Some of their products
Poolside basketball, Pro-Chip Island Golf, Poolside Volleyball, ESPN Swimming challenge, Kickboards, Ocean art light up jellies, Dive sticks, Lazy Dayz lounge, Spring float recliner, Battle Shark ™, Zip Pets™, Fountain Fairies™, Swim diaper, Swim vest,

Baby spring float, Kids canopy chair, Beach canopy chair, and much more.

Idea submission and submission policy
Swimways introduces approximately 20 new products each year. Some are developed internally but many come from outside sources. If you have a great idea, visit their website.

At *home page* click on *Inventor's* link to access their Invention Disclosure Agreement.

In order for your submission to be considered you must have a detailed written description of the invention and drawings (color or black & white). You should retain a duplicate copy of these materials.

Swimways will not examine or accept an idea that is merely outlined in writing or described to them orally.

They may also ask you to submit additional materials of the concept for consideration such as models, working prototypes, design sketches, etc. These additional materials can be returned to you upon request.

Print out and send the disclosure agreement to the above address or email.

After submitting your idea, please do not contact them by phone. You will receive an email from Swimways to update you on whether your idea was accepted or not.

Comments
Swimways sets its product line for a year and a half in advance. As a result, depending on when you submit your idea for consideration, the decision-making timeline may vary.

Don't Get Scammed

"There's a sucker born every minute" and with the advent of the internet that number is getting larger. Even this particular phrase has scammed many, its origination erroneously being credited to PT Barnum.

If you have come up with an invention or product idea you should be cautious about who you approach for assistance. There are a number of disreputable companies that don't follow through on their promises and oftentimes bilk aspiring inventors out of their money. Before signing on with anyone, always do your research.

To check out complaints filed against invention promotion companies, you can go to the Better Business Bureau, or visit www.uspto.gov/web/offices/com/iip/complaints.htm

Thane International, Inc.

What they do
Thane is a direct response marketing company

Stuff they sell
Fitness products, House Wares, Beauty Products

Website
Home page: http://www.thaneinc.com/

Submission page:
http://www.thaneinc.com/submissions.php

HOW TO REACH THEM

Address & Phone numbers
See website for all of Thane's offices. Click on appropriate address for more information.

Email
Use the website contact form.
http://www.thaneinc.com/contact.php
(general inquiries)

Some of their products
Slim 'N Lift Supreme™, Klear Action™ Whitening Light™, CA Beauty® Sudden Glow™, California Beauty® Sudden Lift, and more.

Idea submission and submission policy
If you have an idea, a finished project or a completed direct response campaign, go to submission link above or (1) Go to *home page* (2) Click on *click here for product submissions.*

Product idea must be submitted via their website link.

Comments
Thane will complete submission review of your product within four weeks.

Thermo King

Stuff they sell
Climate Control/Refrigeration Products, Heating and Freeze Protection

Website
Home page: http://www.thermoking.com

HOW TO REACH THEM

Address
Thermo King Corporate Headquarters
314 West 90th Street
Minneapolis, Minnesota 55420

Phone Number
(952) 887-2200 (General inquiries)

Email
bridgeton_contact_center@irco.com

Some of their products
Thermo King Corporation manufactures transport temperature control systems for trailers, truck bodies, buses, shipboard containers and railway cars and more.

Food and beverage vehicles, Insulated thermal panels, Thermostat controls, Truck blowers,

HVAC driver cabin units, HVAC Passenger car units, Mount and drive kits. See site for full list.

Some of their brands
Thermo King Magnum ®, Heat King 400, Heat King 400 HO, and much more. See website for complete list.

Idea submission and submission policy
Thermo King says they are always interested in new product ideas. However, they must follow certain precautionary procedures before accepting disclosures of new ideas to avoid any confusion between parties. Accordingly, they require the signing of an invention submission form before taking further action. Before contacting them these are some things to consider:

- Does the product idea have patent protection?
- Do you want to sell the rights to the product, manufacture it yourself, or license the rights to the product with a royalty?
- Have any market studies been done for such a product?
- Has a patent search for similar products or ideas been done?
- What type of development and test time has the product had?

Use the contact information listed above to request invention submission form.

Thexton

Stuff they sell

Tools for the Automotive Market: Automotive specialty tools are geared specifically to the professional auto mechanic and technician:

Information systems, Brake Tools, Fuel Systems, Specialty Tools, Charging System Test Adapters, Cooling Systems, Repair Parts & Kits, Diagnostic Accessories, and Electrical & Battery

Website

Home page: http://www.thexton.com

Submission information page:
http://thexton.com/index. cfm/preset/ideas

HOW TO REACH THEM

Address

Thexton Manufacturing Company
1157 Valley Park Drive
Suite 150
Shakopee, MN 55379
Attention: Inventions

Ross

Phone Number

Phone: (800) 328-6277
Fax: (952) 831-5938

Email
customerservice@thexton.com

Some of their products
Ford cruise control tester, Full fielder alternator test adapter, Oxygen sensor/fuel circuit tester, Test lead, GM idle air passage plugs, Brake return spring, and much more.

Idea submission and submission policy
Go to submission information page above or (1) Go to *home page* (2) Click on *product idea* link where you can fill out product submission form.

Submissions can be made on Thexton's website, printed out and mailed to the above address, or faxed to the above facsimile number.

Comments
Read website to learn more about Thexton.

3M

Stuff they sell
Displays and Graphics, Electronics, Electrical, Healthcare, Communications, Safety and Security, Transportation Industry, Office, Home and Leisure, Manufacturing and Industry

Website
Home page: http://solutions.3m.com/en_US/

Submission page:
http://solutions.3m.com/wps/portal/3M/en_US/SubmitYourIdea/

HOW TO REACH THEM

Address
3M Corporate Headquarters
3M Center
St. Paul, MN 55144-1000

(Not for submissions)

Phone Number
(888) 364-3577

(Not for submissions)

Email
Email questions via website contact form

Some of their products

Home & Leisure: First Aid, Air cleaning filters, antacid, facial and body sponges, insect repellent, Car care and repair, Paint protection, Personal protective equipment, Scrap bookers glue, Vellum tape.

Office: Spray adhesives, glue sticks, Specialty tapes, Adhesive transfer products, Packing list envelopes, Filament tape,

Displays & Graphics: Touch screen systems, Traffic safety systems, Computer privacy filters, Large format Graphics

Electronics, Electrical & Communication: Electronics design and manufacturing, Communications technologies.

Healthcare: Animal care, Dental, Microbiology, Stethoscopes, Infection prevention.

***Manufacturing & Industry*:** Abrasives, Sand paper, Orbital sanders, Industrial adhesives, 3M™ Temperature Logger TL20.

Safety & Security: Border & document security, Brand security, Anti-slip products, Fire protection products.

Transportation Industry: Automotive aftermarket, Accessory attachment, Automotive OEM's, Marine liquid wax. This is only a small sampling of a very large products and services list. See website for complete information.

Some of their brands

Scotch®, Scotch-Brite®, Command Strips™, Post-It®, Scotchgard™, Nexcare™, Scotchlite™, and more.

Idea submission and submission policy

At home page, go to *Products & Services* link then click on *submit your idea*, or go to submission link listed above for 3M's conditions and complete Patented Idea Submission Form.

3M will only accept patented or patent pending applications for product ideas. No email, phone, or regular mail submissions will be accepted. All submissions must be made through the 3M website.

Tupperware

Stuff they sell
Plastic Kitchen Storage Containers, Kitchen Cookware, Toys, Microwave Cookware

Website
Home page: http://tupperware.com

HOW TO REACH THEM

Address
Tupperware Worldwide
International Law Department
P.O. Box 2353
Orlando, FL 32802-2353
Attn: New Products Coordinator

Phone Number
(800) 366-3800

Email
Contact Us link at site offers an email form

Some of their products
Bowls, Food storage, Spring tumblers, Easy store and pour set, Cake taker set, Toys, Salsa set, Colander, Spatulas, Butter dishes, Heat 'n Serve™ containers, Crystal Wave™ soup mug, Micro pitcher set, Jel-Ring® Mold, Cold cut keeper, Canister sets, Cereal Storer, Sandwich

keeper set, Cutlery, Frying pan, Sauce pan, Cookware set, and much more. See website for all of their products.

Idea submission and submission policy
Contact Tupperware/Dart Industries Inc. and request a Condition of Submission form. After signing this agreement, send this along with your product/idea description, and sketches, drawings, photographs, etc. to the above address.

Comments
Tupperware recommends that your product idea be patented before submitting.

Unilever

Stuff they sell
Food & Beverage, Home & Personal Care

Website
Home page: http://www.unilever.com/

Submission Home page:
http://www.ideas4unilever.com

HOW TO REACH THEM

Address
N/A

Phone Number
N/A

Email
open.innovation@unilever.com
(Any submission questions)

Some of their products
Butter, Mayonnaise, Ice-cream, Dish soap, Mustard, Peanut butter, Toothpaste, Soap, Tea, Soups, Laundry soap, Vaseline, Salad dressing, and much more. See *home page* for more information on products and brands.

Some Unilever owned companies
Bertolli, Hellmans, Fudgesicle, Sunlight soap, Skippy, Colman's Mustard, Sunsilk, Sunlight, Slim Fast, Lipton, Knorr, Country Crock, and more.

Idea submission and submission policy
Go to Submission page above, or, from home page go to *Innovation* link then click on *collaborating with us* link and go to *working with us*.

Follow these steps to submit your non-confidential proposal:

Read the Confidentiality Waiver form. Link is on the web page. Unilever recommends that you seek legal advice if you feel you should or if you don't understand the form.

Print out the Confidentiality Waiver form. Fill out, sign and fax it to Unilever at the number indicated on the form, or e-mail a scanned copy of the signed form to Unilever at open.innovation@unilever.com.

Once you receive a reference number from Unilver, send them your submission marked with the reference number which you will find at the top of the returned copy of the Confidentiality Waiver Form.

Please do not send confidential information before completing a Confidentiality Waiver form. This protects Unilever's rights upon receipt of outside ideas and eliminates the risk of jeopardizing on-going research programs, projects and developing concepts within Unilever.

Do not send any ideas relating to product names, advertising, contests, promotions and trademarks as they are outside the scope of Open Innovation. And please do not send proposals if you are under 18 years of age.

This and further information can be found on submission page.

Comments
Study Unilever brands and know your market. Contact Unilever to learn their current needs.

Unilever also offers venture capital for business start ups. See link under submission policy on submission page for more information.

West Bend

Stuff they sell
Small kitchen appliances:
Blenders, Mixers, Smoothie Makers, Popcorn Poppers, Coffee Makers, Skillets, Griddles, Grills, Toasters, Ovens, Bread Makers, and more.

Website
Home page: http://www.focuselectrics.com/

HOW TO REACH THEM

Address
West Bend
2845 Wingate Street
West Bend, WI 53095

Phone number
(262) 334-6949
Fax: (262) 306-7026

Email
To email go to *contact us* page to send inquiries.

Idea submission and submission policy
Go to *home page* and click on *West Bend* images then click on *FAQ.* Go to, *New Product Submission* link.

Print out and sign non-disclosure form and attach it to the description or disclosure of the idea and mail to the attention of West Bend's Product Development Manager at the above address.

Comments

West Bend will accept submissions under these rules, but they are not as open to new ideas as most of the other companies in this book.

Question:

I have my finished prototype, but now I'm not sure what to do next. I believe I need to let people use my prototype. Is that correct?

Answer:

Getting feedback by allowing your target market to put your prototype to the test is always a good idea. It will help you get a feel for potential demand as well as work out any bugs or possibly improve it with suggestions.

Once you're satisfied that it's ready for the public, and sufficiently protected, you will want to look into having it manufactured. You can choose to have it manufactured and distributed yourself. This obviously will come at a cost to you, but it will also produce bigger profits in the long run. Or you can seek out manufacturers that will buy or license your invention/product. This will usually result in a one time up front payment to you or royalties on each sale, depending on the agreement you and the manufacturer come to.

Wham-O

Stuff they sell
Indoor and Outdoor Recreational Products

Website
http://www.wham-o.com/

HOW TO REACH THEM

Address
Wham-O, Inc.
5903 Christie Ave.
Emerysville, CA 94608

Phone Number
(888) 942-6650

Email
coplin.e@wham-o.com
consumer_affairs@wham-o.com

Some of their products
Sea-Doo® Kayak, Sea-Doo® Ray, Snowboards, Body Boards, Powder Boards, Lawn Fishing ™, Splash Factory ™, Wham-O® Boggan, Frisbee® Horse Shoo, Croquet Golf ™, Glow in the Dark Air Darts ™, Snow Boogie® Sleds, and so much more. See website for complete product list.

Some of their brands
Frisbee®, Sea-Doo®, Morey®, Slip ‘n Slide®, Air Zinger™, Super Ball ®, Hula Hoop ®, Hacky Sack ™

Idea submission and submission policies
Wham-O will not be taking new product submissions again until September 2010. At that time, you can contact Wham-O for a non-disclosure agreement.

Wilson

Stuff they sell
Sports Equipment

Website
Home page: http://www.wilson.com

HOW TO REACH THEM

Address
Wilson Sporting Goods
Attn. Research & Development/(Sport)
8750 W. Bryn Mawr Ave
Chicago, IL 60631

Phone Number
(773) 714-6400 (Questions)

Email
N/A for submissions

Some of their products
Ball gloves, Catchers gear, Baseballs, Shuttlecocks, Badminton racquets, Tennis racquets, Tennis balls, Footwear, Racquetball gloves, Eyewear, Backboard systems, Uniforms, Basketballs, Golf clubs, Golf balls, Cart bags, Carry bags, Headwear, and much more. See website for more products.

Idea submission and submission policy
The Research & Development department asks that all requests be sent in through the mail for review. You do not have to disclose the entire product, just the basic information. Please send information to the address listed above.

Comments
No email submissions.

Wirthco

Stuff they sell
Battery & Fuse Block Accessories, Funnel & Fluid Control Products, Maintenance & Safety Products, Camping Products.

Website
Home Page: http://www.wirthco.com

HOW TO REACH THEM

Address
Wirthco Engineering
7491 Cahill Road
Minneapolis, MN 55439

Phone Number
(952) 941-9073
Toll Free: (800) 959-0879
Fax: (952) 941-0659

Email
Go to the *contact us* link for email form

Some of their products
Ammo & Gun Storage, Soap, Terrariums, Clothesline, Ishy Fishy Soap Dope, Shovels, Eye Shield, Blizzard Bucket Safety Kit, Sand Bags, Rescuer Shovel, Clip on Funnel, EZ Smart Drum Funnel, Funnels that Fold and

much more. See their website for their full product line.

Idea submission and submission policy
Contact Andrew or Steve Wirth at Wirthco Engineering via email to introduce your idea. Andrew@wirthco.com or swirth@wirthco.com

Xerox Corporation

Stuff they develop
Aerospace, Advanced Materials, Automotive, Transportation, Electrical/Electronics, Biomedical/Pharmaceutical Image/Data Processing Chemical/ Specialty Materials, Mechanical, Communications, Nanotechnology, Consumer Products, Software Electronics, Financial Services, Graphic Arts/ Imaging, Industrial Products, Manufacturing, Optics, Security

Website
Home page: http://www.xerox.com/
(product information found at this site)

Submission home page:
http://xeroxtechnology.com/

HOW TO REACH THEM

Address
Xerox Corporation
0139-21A
800 Phillips Road
Webster, NY 14580-9720

(Not for submissions)

Phone Number
Not available for submissions

Email
Not available for submissions

Some of their products
Multifunction printers, Copiers, Software, Scanners, Disc duplicators, Digital presses, Production printers & copiers, Continuous feed printers, Wide format, Work-flow and software. See submission *home page* for their list of technologies and industry.

Some of their brands
Carbon ConX ™

Idea submission and submission policy
By policy, Xerox requires that you complete their standard non-confidential disclosure agreement if your submission includes anything other than issued patents. This agreement states that no confidential relationship or obligation of secrecy is created between the submitter and Xerox Corporation by the submission and its consideration by our company.

Xerox policy requires all submissions be only in writing (including figures, pictures, and drawings). Xerox does not have face-to-face meetings with submitters or their representatives in evaluating unsolicited submissions.

To submit: (1) Go to submission *home page* (2) Click on *submission* link at the bottom of the page.

Instructions will be listed along with a link for website submission of your idea.

Comments
Xerox encourages and welcomes unsolicited ideas and suggestions. They recommend that you protect your submission.

Zebco

Stuff they sell
Fishing Products: Rods, Reels, Combinations, and more

Website
Home page: http://www.zebco.com/

Submission page:
http://www.zebco.com/common/ideas.html

HOW TO REACH THEM

Address
Zebco
6105 E. Apache
Tulsa, OK 74115

Phone Number
(800) 444-5581
(Comments and questions)

Email
N/A

Some of their products and brands
Zebco Slab seeker®, Slab Seeker combos, Slab Seeker Rods, Zebco Hawg Seeker™, Bite alert spin cast, Zebco Pro Staff®, Zebco Trout Seeker™, Bait action triggerspin, Zebco

Platinum™, Zebco Omega™, Tackle wallet, Kids fishing products and much more.

Idea submission and submission policy
Zebco's policy is as follows:
Zebco appreciates the fact that you have considered them for the submission of an idea. However, to avoid potential misunderstandings or disputes it is Zebco's policy ***NOT*** to accept any unsolicited ideas, innovations, product enhancements, new promotions, new advertising or marketing campaigns, products, technologies, processes, materials, or ideas for new product names.

Zebco ***WILL***, however, accept ideas for consideration for which a valid patent has been issued and for which the owner of the patent is submitting the idea.

Please know that, despite their request that you only submit ideas for which a valid patent has been issued, any unsolicited letter or package that is sent to Zebco that is identifiable as an unpatented idea will be discarded unopened. No email submissions. All e-mails will be deleted.

Zebco can make no assurances that your ideas and materials will be treated as confidential or proprietary.

If you submit an idea for which you have a valid patent, please write the patent number on the outside of the package. Patent pending ideas will not be considered.

Interview with inventor, Baruch Breuer

Mr. Breuer (http://www.waterstop-loss.com/) has invented a water saving device that recycles what is called gray-water. Gray-water is the waste water from showers and sinks that can be put to other uses.

How effective is your invention in water recycling?

Very effective. One to two showers per day provide about 100-150L of gray-water, enough for all toilet flushings (15 -20 times X 6L) per day so that the clean water is almost never used.

How does the process work?

A small 8 mm diameter, flexible rubber pipe and electric sensor is inserted into sink/shower/tub/washing drain, to draw the gray-water out with small electric pump and is filtered and recycled into a small tank without any changes to the drain pipes or flush-tank.

If gray-water is not sufficient or available the system automatically switches to clean water.

How easily is the device installed? Does it require any special expertise?

No special expertise is required once the determination is made where the best collection source for gray-water is located.

The electric pump and sensor must have 110/240 connection.

What can the recycled water be used for?

Flushing toilets or garden watering.

How frequently would the unit need to be replaced?

To avoid odors, gray-water should be disinfected and monitored periodically. Standard toilet/pool disinfectants are used and changed 1-3 monthly.

Is this your first invention?

Yes, in gray-water recycling.

Anything else you would like to share?

I do not have the time or means to promote this product on large scale, but am very happy with few installations already made.

I would be happy to cooperate with anyone who genuinely believes in this cause and wants to develop it further.

PART TWO

Businesses, Organizations & Publications for the Inventor

Entrepreneur Magazine

Entrepreneur's e-zine and hardcopy magazine provides information on business opportunities, home based businesses, franchises, financial direction and more.

Website

http://www.entrepreneur.com/

Inc.

Inc. is an online and hardcopy magazine with content that covers a variety of entrepreneurial, innovative and business directed subjects.

Website

http://www.inc.com/

Job Shop

Find manufacturers for your custom contract needs. The Job Shop hooks up buyers, engineers and vendors to outsource custom parts, services, and assemblies in North America.

Website

http://www.jobshop.com/

National Inventor Fraud Center

Founder, Michael S. Neustel, is a U.S. Registered Patent Attorney. The purpose of NIFC is to provide information to consumers about invention promotion companies and how people can market their ideas. National Inventor Fraud Center is owned by Neustel Law Offices, LTD.

Website
http://www.inventorfraud.com/

MFG.COM

MFG connects businesses with quality suppliers that have the expertise to meet sourcing needs.

Website
Mfg.com

IP Watchdog

IP Watchdog provides free resources on intellectual property law and related topics.

Website
http://www.ipwatchdog.com/

Trademark Express

Trademark Express researches and files trademark applications. Their services also include patent, logo, copyright, fictitious name research, and more. See their website for more information.

Website

http://www.tmexpress.com/

TSW (Trade Show Week)

Website with tradeshow schedules

Website

http://www.tradeshowweek.com/

United Inventors Association

The United Inventors Association (UIA) is the world's leading center for inventor/ entrepreneur education. UIA networking and support extends throughout the USA and abroad.

Website

http://www.uiausa.org/

USPTO

The USPTO (United States Patent and Trademark Office) will answer your questions on patents, trademarks and copyrights, allow you to do patent, trademark, and other searches and is set up for you to file online.

Website
http://www.uspto.gov/

The Launch Hour

Launch is a weekly radio show for inventors and consumer product entrepreneurs to pick up tips and get professional advice from those who have gone before. Tune into their podcasts and check it out.

Website
http://thelaunchhour.businessradiox.com

Invention Radio

Have questions? Get expert answers and advice with this radio program for entrepreneurs and inventors.

Website
http://www.gotinvention.com/main.php

Emory Day

Commission based, with a small commitment fee, Emory Day helps businesses, entrepreneurs, and inventors sell their products with professional web design, search engine, social media, e-commerce, affiliate marketing and publicity. You can choose the full package or specific areas you need assistance in.

Website
http://emoryday.com/

Mom Invented

Mom Invented assists moms with business start ups and product/invention development. They also schedule product searches for their *Mom Invented® brand* product line.

Website
http://www.mominventors.com/

Simple Patents

Patent searches, Patent applications.

Website
http://www.simplepatents.com/

KimCo

KimCO is in business to help companies with product development, prototyping and manufacturing overseas. Because of their extensive connections with retail buyers, they can also assist with placing products in multiple outlets, including QVC, Wal-Mart, Target and Home Depot.

Website

http://www.kimbabjak.com

ThomasNet

Search and connect with manufacturers and wholesale suppliers for your business needs.

Website:

http://www.thomasnet.com/

PART THREE
Licensing, Copyrights, Patents & Pitching Your Idea

Basic Licensing Facts

(Definition of Intellectual property: Property that can be protected under federal law, including copyrightable works, ideas, discoveries, and inventions.)

When you license your intellectual property it grants another individual or business the right to make use of it in the manner agreed upon. The owner/creator of the intellectual property is the licensor and the license holder is the licensee.

You can choose to license your work to one entity or several at a time. A *non-exclusive license* allows the intellectual property rights to be granted to more than one licensee.

An exclusive license can be defined as a license granted to a single licensee, but it can also mean that the license is exclusive to certain geographical areas, product areas, or to a limited area of use.

The type of license you need depends upon the intellectual property being licensed:

A patent license grants permission to use or sell a patented product, design, or process.

A *trademark or service mark license* grants permission to the owner to allow another individual or entity to use the mark. If you have a trademark or service mark you are considering licensing, you need to be aware that as the licensor you have a responsibility to see that the source and quality of the goods and services produced and sold by the licensee are equal to the trademark or service mark name you have established.

A *know-how* license is usually used to transfer the licensor's technological knowledge in a particular area or field for use by the licensee.

A *trade secret license* grants permission to the licensee to make, use or sell a product, design, or process using a trade secret. Coca-Cola's recipe for their legendary soft drink, or KFC's secret blend of herbs and seasonings would be considered trade secrets.

Copyright licensing or assignation grants permission to use literary, artistic or musical forms of creative expression developed by the licensor. This includes, but is not limited to novels, poetry, architecture, artwork, and software.

When you are ready to pursue a licensing agreement with a manufacturer, your first step is to locate manufacturers that make the same or similar type of product as yours. Some manufacturers are interested in product ideas developed outside of their own research and development departments and others are not.

You can make multiple submissions at one time, contacting several manufacturers or inventor agents to pitch your idea to.

Patent Q & A

What is a patent?
A patent is a property right that is granted by the government to the inventor. A patent does not grant the right to make, use, or sell an invention, but the right to prevent others from making, using or selling it.

What can be patented?
According to the language of the statute, any person who "invents or discovers any new and useful process, machine, manufacture, or composition of matter or any new and useful improvements thereof, may obtain a patent."

Are patents necessary?
If your invention falls under any of the above categories and you want the legal assurance that your invention remains yours, then it is necessary. Your invention does not have to have a patent but you do put yourself at risk and have less chance of successfully suing if someone steals your idea.

What is a utility patent?
A utility patent is the most commonly used and lasts for 20 years from the date of publication. There are fees charged to the patent holder at 3 ½ years, 7 ½ years and 11 ½ years to maintain the patent's protection.

What is a design patent?
Design patents are granted to anyone that has invented a new, original and ornamental design for an article of manufacture. This type of patent protects the appearance of the article but not its structure or function. They are good for 14 years from the date the patent is granted and there are no maintenance fees.

What is a provisional patent?
Cheaper than a standard utility patent, provisional patents protect your invention for one year at a lesser cost than the traditional patent. This will allow you to shop your idea around to potential manufacturers or safely manufacture it yourself and produce a profit before investing in the cost of a utility patent.

Applications for provisional patents must include the names of all involved inventors and can be filed up to 12 months following the date of first sale, offer for sale, public use, or publication of the invention.

Is a poor man's patent legally acceptable?
A poor man's patent is the practice of sealing up your invention idea in an envelope and mailing it to yourself. The idea is to keep it sealed and use the stamped date on the envelope as proof of when your idea was protected. Unfortunately this process is ineffective when it comes to defending or proving ownership

Can I file a patent myself?
Anyone can file for a patent. However, to make sure it is done correctly, you would be better off hiring a legal entity to do it for you. Most will also do patent searches. A search is an essential step before filing to make sure there is not an invention like yours already patented.

How much do patents cost?
The filing fees vary but current rates can be found by going to http://www.uspto.gov

Copyright

What does copyright law cover? Copyright protects the creator's specific expression in literary, artistic, or musical form. This ranges from software, movies, screenplays, novels, artwork, poetry, music, architecture and other works in these areas. It does not cover any idea, system, method, device, name or title. These fall under patents and the latter under trademarks.

More specifically, under the visual arts aspect, copyright covers pictorial, graphic or sculptural works, including two and three dimensional works of fine, graphic, and applied art, photographs, prints and art reproductions, maps, globes, charts, technical drawings, diagrams, architectural works and models.

Your work is legally yours and under copyright protection from the moment it is created and in a tangible form. A formal registration of copyright is not necessary unless you intend to file a lawsuit for infringement. However, registered works may also be eligible for statutory damages and attorney fees in the event of a lawsuit.

If registration is submitted within 5 years of publication your copyright will be considered solid evidence to establish rightful ownership.

You can pay someone to register a copyright for you, but they are relatively simple and very inexpensive to file yourself.

To file for copyright protection go to: http://www.copyright.gov/

Pitching Your Idea

When you're ready to take the leap and sell your idea to manufacturers or licensing agents, you are no longer an inventor but a salesman. In your pitch to the manufacturer be it verbally or in writing, you want to provide solid evidence that your product will sell well.

For the sake of persuasion, use details about how your product will fill a consumer need. If there are similar products already available, describe in what ways yours is better or different. Know your product's qualities and benefits and don't be shy about praising them.

You'll also want to describe how your product idea can make people's lives easier or better. Will it save them money? Make them healthier? Bring pleasure to their lives? These are important selling points that will enhance your presentation and hopefully gain and keep the attention of the decision maker.

If your idea has the potential to be used in ways other than its basic form, or it can be expanded or altered by including add-on products, use this as a sales feature when you pitch your idea.

A simplified, non-technological example of this would be the Barbie doll. Barbie is your product idea and you have the initial market for the doll, but Barbie is hip and needs a wardrobe so you make clothing. Now the plastic perfection with the one inch waist and pre-formed stiletto feet needs a better half to take her shopping for those clothes so you introduce Ken...And what better way to get to the mall than to hop in the hot pink Barbie car and drive there?

Clothes, boyfriend doll, car, these are all additional products or add-ons that can be used to enhance the selling of your initial product.

Prior to submitting your information, manufacturers and agents will specify what it is they want from you and at what point they want it. At the initial stage, some may request a simple sketch and/or a basic description of what your invention or product does and how it works. If your idea is patented, sending the patent information is sometimes all that is required. However, if sending additional information is not prohibited in the rules for submission, then go one step further and include a simple brochure or printout with pictures or sketches and information that promotes the virtues of your product.

Sell your product idea with confidence, showing that you believe in it enough to have done the research. When you make your pitch, be professional, follow the submission guidelines explicitly and use whatever facts, market statistics or comparisons you have at your disposal to support the excellence of your idea.

CPSIA information can be obtained at www.ICGtesting.com
Printed in the USA
LVOW121152050212

267147LV00015B/81/P